NATIONAL ACADEMIES *Sciences Engineering Medicine*

NATIONAL
ACADEMIES
PRESS
Washington, DC

Planning the Future Space Weather Operations and Research Infrastructure

Committee on Space Weather Operations
and Research Infrastructure Workshop,
Phase II

Space Studies Board

Division on Engineering and Physical
Sciences

Proceedings of the Phase II Workshop

THE NATIONAL ACADEMIES PRESS 500 Fifth Street, NW Washington, DC 20001

This study is based on work supported by Contract NNH17CB02B with the National Aeronautics and Space Administration and Grant 2126142 with the National Science Foundation. Any opinions, findings, conclusions, or recommendations expressed in this publication do not necessarily reflect the views of any agency or organization that provided support for the project.

International Standard Book Number-13: 978-0-309-69366-0
International Standard Book Number-10: 0-309-69366-7
Digital Object Identifier: https://doi.org/10.17226/26712

Copies of this publication are available free of charge from

Space Studies Board
National Academies of Sciences, Engineering, and Medicine
Keck Center of the National Academies
500 Fifth Street, NW
Washington, DC 20001

This publication is available from the National Academies Press, 500 Fifth Street, NW, Keck 360, Washington, DC 20001; (800) 624-6242 or (202) 334-3313; http://www.nap.edu.

Printed in the United States of America.

Suggested citation: National Academies of Sciences, Engineering, and Medicine. 2022. *Planning the Future Space Weather Operations and Research Infrastructure: Proceedings of the Phase II Workshop.* Washington, DC: The National Academies Press. https://doi.org/10.17226/26712.

COMMITTEE ON SPACE WEATHER OPERATIONS AND RESEARCH INFRASTRUCTURE WORKSHOP, PHASE II

CHRISTINA M.S. COHEN, California Institute of Technology, *Co-Chair*
TUIJA I. PULKKINEN, NAS, University of Michigan, *Co-Chair*
DANIEL N. BAKER, NAE, University of Colorado Boulder
ANTHEA J. COSTER, Massachusetts Institute of Technology Haystack Observatory
MARY K. HUDSON, Dartmouth College
DELORES KNIPP, University of Colorado Boulder
KD LEKA, NorthWest Research Associates
CHARLES D. NORTON, NASA Jet Propulsion Laboratory
TERRANCE G. ONSAGER, National Oceanic and Atmospheric Administration/Space Weather Prediction Center
LARRY J. PAXTON, Johns Hopkins University Applied Physics Laboratory
PETE RILEY, Predictive Science Inc.
RONALD E. TURNER, Analytical Services, Inc.
NICHOLEEN M. VIALL-KEPKO, NASA Goddard Space Flight Center
ENDAWOKE YIZENGAW, The Aerospace Corporation

STAFF

ARTHUR CHARO, Senior Program Officer, *Study Director*
ALEXANDER BELLES, Christine Mirzayan Science and Technology Policy Graduate Fellow
GAYBRIELLE HOLBERT, Senior Program Assistant, Space Studies Board
COLLEEN N. HARTMAN, Director, Space Studies Board

Preface

Space weather has been described as "including any and all conditions and events on the sun, in the solar wind, in near-Earth space and in our upper atmosphere that can affect space-borne and ground-based technological systems and through these, human life and endeavor."[1] Affecting technological systems at a global-scale, space weather can disrupt high-frequency radio signals, satellite-based communications, navigational satellite positioning and timing signals, spacecraft operations, and electric power delivery with cascading socioeconomic effects resulting from these disruptions. Space weather can also present an increased health risk for astronauts, as well as aviation flight crews and passengers on transpolar flights.[2]

Recent executive and legislative efforts have worked toward improving U.S. preparedness for space weather events and strengthening the infrastructure vital to national security and the economy. Notably, in March 2019, the White House released the National Space Weather Strategy and Action Plan (NSW-SAP), which contained strategic objectives and actions necessary to achieve a space weather–ready nation and provided guidance for activities of the Space Weather Operations, Research, and Mitigation (SWORM).[3] Passage in December 2020 of Public Law 116-181, the Promoting Research and Observations of Space Weather to Improve the Forecasting of Tomorrow Act (PROSWIFT Act), codified the elements of the NSW-SAP, prescribed the roles and objectives of the relevant federal agencies, and directed efforts to improve the transition from research to operations. All these activities speak to the importance of timely and accurate space weather forecasts.

In 2019, the National Academies was approached by the National Aeronautics and Space Administration (NASA), the National Oceanic and Atmospheric Administration (NOAA), and the National Science Foundation (NSF) to organize a workshop that would examine the operational and research infrastructure

---

[1] National Aeronautics and Space Administration, 2014, 11. What Is Space Weather, In "Solar Storm and Space Weather - Frequently Asked Questions," https://www.nasa.gov/mission_pages/sunearth/spaceweather/index.html#q5.

[2] American Meteorological Society, 2013, "Space Weather," https://www.ametsoc.org/index.cfm/ams/about-ams/ams-statements/statements-of-the-ams-in-force/space-weather.

[3] In 2016, an executive order created the SWORM Subcommittee under the auspices of the National Science and Technology Council (NSTC) of the White House Office of Science and Technology Policy (OSTP) in order to coordinate efforts across the federal government regarding space weather.

that supports the space weather enterprise, including an analysis of existing and potential future measurement gaps and opportunities for future enhancements. This request was subsequently modified to include two workshops, the first (Phase I) of which occurred in two parts on June 16-17 and September 9-11, 2020. A proceedings summarizing that workshop was published in 2021.[4]

The Phase I workshop was sponsored by NOAA, in consultation with NASA and NSF. The task statement for the workshop focused on space weather operations, including measurement continuity needs. Following the 2020 workshop, NASA and NSF, in consultation with NOAA, requested a follow-on workshop (Phase II) that would focus on the research agenda and observations needed to improve understanding of Sun-Earth interactions that cause space weather. Specifically, the Phase II workshop organizing committee was asked to

1. Examine trends in available and anticipated observations, including the use of constellations of small satellites, hosted payloads, ground-based systems, international collaborations and data buys, that are likely to drive future space weather architectures; review existing and developing technologies for both research and observations;
2. Consider the adequacy and uses of existing relevant programs across the agencies, including NASA's Living With a Star (LWS) program and its Space Weather Science Application initiative, NSF's Geospace research programs, and NOAA's Research to Operations (R2O) and Operations to Research (O2R) programs for reaching the goals described above;
3. Consider needs, gaps, and opportunities in space weather modeling and validation, including a review of the status of data assimilation and ensemble approaches;
4. Consider how to incorporate data from NASA missions that are "one-off" or otherwise non-operational into operational environments, and assess the value and need for real-time data (e.g., by providing "beacons" on NASA research missions) to improve forecasting models; and
5. Take into account the results of studies, including NASA's space weather science gap analysis (part of the NASA Heliophysics Division's Space Weather Science Application program) and the NSF report *Investments in Critical Capabilities for Geospace Science* (2016), to identify the key elements needed to establish a robust research infrastructure.

The Phase II workshop, conducted virtually due to COVID-19 travel restrictions, occurred on April 11-14, 2022, with sessions on agency updates, research needs, data science, observational and modeling needs, and emerging architectures relevant to the space weather research community and with ties to operational needs. The presentations, posters, and videos can be found on the project website, Space Weather Operations and Research Infrastructure Workshop, Phase II.[5] Robert Pool served as a rapporteur for the workshop and provided the committee with an initial draft of the proceedings.

The workshop organizing committee planned the sessions by creating a list of questions that would guide the invited speakers to address the workshop task statement issues. Each session was led by a member of the organizing committee. The present proceedings is organized around the workshop agenda (Appendix B), modified for clarity and conciseness. Each chapter starts with a summary of major themes emerging from the workshop presentations and discussions that are relevant to the statement of task. It should be noted that the workshop speakers did not provide direct answers to the task element on the

---

[4] National Academies of Sciences, Engineering, and Medicine, 2021, *Planning the Future Space Weather Operations and Research Infrastructure: Proceedings of a Workshop*, Washington, DC: The National Academies Press, https://doi.org/10.17226/26128.

[5] National Academies of Sciences, Engineering, and Medicine, "Space Weather Operations and Research Infrastructure Workshop, Phase II," https://www.nationalacademies.org/our-work/space-weather-operations-and-research-infrastructure-workshop-phase-ii, accessed July 8, 2022.

use of "one-off" missions for space weather operation; however, several participants expressed the need for advance cross-mission planning and (international) collaboration in building the infrastructure. Finally, this proceedings includes a summary of discussions on workforce development, which was not explicitly mentioned in the statement of task but was included as "human infrastructure need."

Chapter 1, which aligns with the topics in Session 1, provides context and background for the workshop. Recognizing the different challenges and needs in research, observation, and modeling in the three key research areas, the Sun and heliosphere, the magnetosphere–ionosphere–thermosphere–mesosphere system, and the Ground Effects, respectively, Chapters 2, 3, and 4 focus on each of these areas separately. The final two chapters cover the entire Sun–Earth system in discussing modeling and validation needs of the space weather research community (Chapter 5), and the future architectures needed to establish a robust research infrastructure (Chapter 6).

Addressing future capabilities implies a time scale. Most of the workshop presentations and discussions focused on already existing or shortly upcoming capabilities and assets, as well as the impact of space weather on near-term technological developments (e.g., the impact on systems that will increasingly rely on precision navigation and timing). Given the focus of the discussions, this proceedings covers a time scale of perhaps the next 5-7 years.

Finally, it is to be noted that, following National Academies' practices, this proceedings of a workshop does not include findings or recommendations. While the workshop organizers sought to get the widest possible representation, the contents of this proceedings are limited to the discussions and presentations given.

Reviewers

This Proceedings of a Workshop was reviewed in draft form by individuals chosen for their diverse perspectives and technical expertise. The purpose of this independent review is to provide candid and critical comments that will assist the National Academies of Sciences, Engineering, and Medicine in making each published proceedings as sound as possible and to ensure that it meets the institutional standards for quality, objectivity, evidence, and responsiveness to the charge. The review comments and draft manuscript remain confidential to protect the integrity of the process.

We thank the following individuals for their review of this proceedings:

Seebany Datta-Barua, Illinois Institute of Technology,
Christine Gabrielse, The Aerospace Corporation,
Joe Giacalone, University of Arizona,
David Hysell, Cornell University,
Susan Lepri, University of Michigan, and
Harlan Spence, University of New Hampshire.

Although the reviewers listed above provided many constructive comments and suggestions, they were not asked to endorse the content of the proceedings nor did they see the final draft before its release. The review of this proceedings was overseen by Louis J. Lanzerotti, NAE, New Jersey Institute of Technology. He was responsible for making certain that an independent examination of this proceedings was carried out in accordance with standards of the National Academies and that all review comments were carefully considered. Responsibility for the final content rests entirely with the authoring committee and the National Academies.

Contents

Summary

In 2019, the National Academies of Sciences, Engineering, and Medicine were approached by officials from the National Environmental Satellite, Data, and Information Service (NESDIS) of the National Oceanic and Atmospheric Administration (NOAA), the Heliophysics Division of the National Aeronautics and Space Administration (NASA), and the Geospace Section at the National Science Foundation (NSF) to organize a workshop that would examine the operational and research "infrastructure" that supports the space weather enterprise, including an analysis of existing and potential future measurement gaps and opportunities for future enhancements. That workshop (Phase I) focused on space weather operations, including measurement continuity needs, and a proceedings summarizing the workshop was published in 2021.[1]

Subsequently, NASA, NOAA, and NSF requested a follow-on workshop that would focus on the research agenda and observations needed to improve scientific understanding of the Sun–Earth interactions that cause space weather. A summary of the workshop is presented in this proceedings.

CURRENT SPACE WEATHER EFFORTS

The U.S. space weather community includes a number of stakeholders both in and outside the federal government. NASA is currently establishing a Space Weather Program within its Heliophysics Division. This reflects the newly broadened role of Heliophysics in the research-to-operations/operations-to-research (R2O2R) process, providing space weather information to lunar and planetary endeavors as well as developing new applications for impact mitigation. For example, the HERMES instrument package to be placed on the lunar-orbiting Gateway will make space weather measurements to support lunar operations and demonstrate technologies needed to conduct a human mission to Mars. The Heliophysics System Observatory (HSO) provides observations for basic and applications-focused research as well as real-time operations.

NSF is carrying out a variety of space weather efforts housed in multiple NSF directorates; for example, the Directorate of Geosciences supports the Coupling, Energetics, and Dynamics of Atmospheric

---

[1] National Academies of Sciences, Engineering, and Medicine, 2021, *Planning the Future Space Weather Operations and Research Infrastructure: Proceedings of a Workshop*, Washington, DC: The National Academies Press, https://doi.org/10.17226/26128.

Regions (CEDAR); Geospace Environment Modeling (GEM); and Solar, Heliospheric, and Interplanetary Environment (SHINE) programs as well as facilities such as the National Center for Atmospheric Research (NCAR). The Directorate for Mathematical and Physical Sciences supports the National Solar Observatory, and the Division of Physics supports plasma physics research. NSF also supports a wide network of observing facilities, including solar telescopes, magnetometer networks, radars, neutron monitors, and networks of citizen-science measurements. Although the NSF-supported observing network is primarily ground-based, NSF also supports balloon-based instrumentation as well as CubeSats and public–private partnerships with the space industry. An important priority for NSF is the development of the space science workforce.

The Department of Defense Space Weather Program has recently been reorganized within the U.S. Space Force under Space Domain Awareness. The goal of that effort is to create a holistic picture of the environment within which operations are conducted, with the inclusion of specific system impacts. Reaching that goal will involve development of an integrated software suite consisting of data, models, and applications as well as interagency agreements, which are being established to facilitate sharing of data between agencies, using open architectures, and enabling all participating agencies to improve their capabilities (e.g., those concerning commercial space traffic management).[2]

The NOAA space weather charter complements those of the other agencies and focuses on capacity building to advance space weather policy, services, research, and operations. NOAA's plans for improvement of space weather services involves (1) sustaining fundamental observations, (2) providing accurate models and forecast products, (3) transitioning scientific and technological advances into operations, and (4) supporting the private sector to fill data and technology gaps and to provide value-added services. NOAA's new operations paradigm combines the establishment of dedicated operational observing systems with modeling and service improvements to provide forecasts, warnings, and data. Furthermore, space weather has recently become a third pillar in NOAA's NESDIS structure, which aims to provide an integrated, digital understanding of Earth's environment.

NOAA's long-term support for NASA's crewed missions was continued by the recently signed NASA–NOAA Interagency Agreement on providing space radiation environment support to all human spaceflight missions, including the International Space Station and Lunar (Artemis) and Martian crewed missions. NOAA's Office of Space Commerce is prototyping an Open-Architecture Data Repository (OADR) environment to facilitate research-to-operations (R2O) efforts related to space traffic management.

A central element of the U.S. national space weather effort is the National Space Weather Strategy and Action Plan (NSW-SAP), which was put in place in 2019. U.S. space weather activities are overseen by the Space Weather Operations, Research, and Mitigation Subcommittee in the Office of Science and Technology Policy. The PROSWIFT Act, signed into law on October 21, 2020, codifies U.S. policy to prepare and protect against the social and economic impacts of space weather. It mandated the establishment of a Space Weather Interagency Working Group to coordinate executive branch actions to improve understanding and prediction of space weather phenomena, and a Space Weather Advisory Group to receive advice from academia, the commercial sector, and space weather end users. In addition, the National Academies' Space Weather Roundtable and NASA's Space Weather Council help support communication and coordination among the agencies in the space weather effort.

Priorities in updating the 2019 NSW-SAP include the R2O2R Framework, benchmarks, activity scales, hazard mapping, human exploration, space situational awareness, aviation, and continuity of satellite observations. The intent expressed in the NSW-SAP is to provide a formal interagency structure to ensure an effective space weather R2O2R process.

---

[2] This report frequently references the term, "space traffic management"; however, other terms in common use are "space traffic awareness" and "space traffic coordination."

The U.S. commercial sector, particularly electric power and aviation industries, has a long history of using space weather observations and forecasts in decision making. For example, the North American Electric Reliability Corporation (NERC) has developed mandatory standards for grid operators for risk assessment and mitigation, especially concerning extreme events, thus guiding old and new operators in the field. Space industry, on the other hand, is booming and bringing new actors to the field, which has created a need to educate new satellite providers and operators on how to adapt to and mitigate space weather hazards. For most space weather applications, having the longest possible forecast lead time, ideally 72 hours in advance, is of key importance. However, there are significant challenges in reaching this objective, including the availability of the required observations, the level of scientific understanding of processes that underlie space weather phenomena, and the intrinsic time scales of the processes themselves.[3]

RESEARCH NEEDS

Space weather lies at the intersection of the natural space environment (object of fundamental scientific research) and hazards to human systems and technology (addressed by operational communities interested in engineering risk mitigation). This intersection includes applied science and engineering as well as R2O activities.

Space weather impacts can be broken into five categories: (1) hazards affecting space-borne technology caused by direct drivers from the Sun (e.g., solar flares and radio bursts); (2) radiation effects (e.g., on technology and humans aboard spacecraft and aircraft); (3) ionospheric effects (e.g., affecting satellite communication, navigation, and high-frequency [HF] radio communications); (4) thermospheric expansion (orbital changes and collision hazards of satellites and other objects); and (5) geomagnetically induced currents (GICs) (which can lead to damage to power systems on the ground).

Reaching the level of understanding required to predict geospace system behavior requires a systems approach involving coordinated and concurrent space-borne and ground-based measurements combined with advanced data science, modernized data infrastructure, and modeling tools. For example, in order to better understand the behavior of coronal mass ejections (one of the key drivers of space weather), it is important to track their evolution in real time and in three dimensions as they travel toward and impact Earth. However, such level of coordination is not reflected in the current infrastructure: While past and current NASA space missions have been combined into the Heliophysics System Observatory, this system has not been strategically planned from a systems-science perspective. As each new mission is evaluated on its individual science goals, any systems science approaches have been and will be on an ad hoc basis. Furthermore, the ground-based measurements suffer from a similar lack of coordination. Since systems science requires an extensive, simultaneously operational fleet, it would greatly benefit from international collaboration.

Multi-satellite missions are necessary for understanding the processes that take place over vast regions of space at many different scales, and technological advances now enable missions to study conditions and processes from the smallest scale to mesoscales and the entire magnetosphere–ionosphere–thermosphere system.[4] However, such missions require substantial investments in inter-calibration of the observing instruments as well as developing robust methods to incorporate data into interpretive models. Constellations covering a range of scales are important for understanding processes that extend from the solar atmosphere

---

[3] T. Pulkkinen, 2007, "Space Weather: Terrestrial Perspective," *Living Reviews Solar Physics* 4:1, https://doi.org/10.12942/lrsp-2007-1.

[4] Note that the small scale and mesoscale definitions are region-dependent with the ionospheric small to mesoscales being from tens of meters to 1,000 km, and magnetospheric scales ranging from tens of kilometers to a few Earth radii.

and inner heliosphere to the near-Earth space environment. Similarly, it was noted by some participants that a dense network of ground-based and space-based geophysical observatories, including those operated by the private sector, will be required to provide high-confidence, long lead-time predictions of GICs and other hazards to power systems and other technological assets.

Extreme events are rare, and their impacts are neither well understood nor well represented by existing models. Currently available data sets, which are dominated by quiet space weather, pose challenges for machine learning models as well as validation of physics-based models when they try to address very high activity periods. Continuous monitoring of space weather is needed to observe rare space weather events. However, observation of such events was said to provide a unique opportunity to increase knowledge of their impacts on the space environment, on the upper atmosphere, and on the ground- and space-based technological systems.

These considerations point to the following detailed research needs:

- *Tracking the entire system*: The NASA gap analysis report lists top observational priorities: solar and solar wind observations, including observations off the Sun–Earth line and the solar wind near Earth; plasma sheet electron and ion injections/bursts from the magnetotail into geosynchronous orbit and medium earth orbit regions; ring current and radiation belt electron acceleration processes; ionospheric and thermospheric key parameters; and ionospheric D- and E-region energetic particle precipitation as well as E- and F-region cusp and auroral region precipitation.

- *Understanding global couplings*: Significant efforts are being devoted to finding ways to couple regimes (e.g., solar wind to magnetosphere), physical transitions (e.g., plasma-dominated versus magnetic-pressure-dominated plasmas, high versus low Alfvén speed plasmas, collisional versus collisionless plasmas), and spatial scales (e.g., magnetohydrodynamic versus kinetic plasmas). Such multi-scale and multi-step processes are important in understanding the physics of space weather, but they are not well observed or well modeled. This is especially true for the ionosphere, which is coupled to the magnetosphere and solar wind from above, but also to the thermosphere and even the stratosphere from below; the forcing from below is an important, but often neglected, aspect of space weather.

- *Real-time monitoring*: Such monitoring is critical for both nowcasting and forecasting, but it creates extra challenges for data downlinking systems and capacity and also places additional requirements on research-focused missions, which generally are not concerned with real-time monitoring.

- *Validating models*: Observing system simulation experiments (OSSEs) can be invaluable in providing cost–benefit analyses of adding additional measurements to improve model performance. In addition, OSSEs can be conducted to identify missing physics and to improve the assessment of model errors and error sources.

- *Reaching out to neighboring fields*: For example, collaboration with plasma physics researchers, data scientists, and Earth scientists would be beneficial for gaining new numerical and computational tools to space weather science.

MODELING AND VALIDATION

Current operational space weather models include large, physics-based models that treat the system elements from the Sun to the upper atmosphere and ground. Experiences from the terrestrial weather community indicate that the way to improve the accuracy of predictions is to increase model resolution (computational power), improve representation of physical processes (scientific understanding), and use advanced data assimilation techniques (new analysis methods).

Data needs are still pressing, because—despite the large data sets already acquired—the data are often not well matched to modeling needs. The lack of suitable data arises either because of the quality or format of available data or because the models need measurements that are not made at all, are not made at the right locations, or are not made at sufficient cadence. Since even large data sets contain only a small number of extreme events, the lack of data remains a challenge for many of the new data analysis and model-development methodologies being employed. In particular, the space weather community does not yet have the data needed to create standardized data products suited for machine learning applications. As the need is to acquire simultaneous observations from as many locations as possible, data buys from private companies operating instruments onboard commercial satellites offer a potential solution for some of the data sparsity issues, assuming that the required instruments and data products would be available.

The space weather community is becoming increasingly multidisciplinary, as optimal use of new methods, such as ensemble modeling, data assimilation, or machine learning, requires bringing in experts from computer science, Earth sciences, space technology, and other fields. While these methods have been applied in other fields to other data sets, their application to space weather prediction problems requires development of dedicated data sets as well as methods tuned to the specific problem (i.e., substantial cross-disciplinary learning between the discipline scientists and the methodology developers). Furthermore, due to the large system size, ensemble models, data assimilation, and machine learning applications often need computing and data storage resources that push the current capacity of even the largest supercomputing facilities.

In particular, space weather must address the following modeling and validation issues:

- *Ensemble modeling, data assimilation, and machine learning* are methodologies that have the potential to significantly improve space weather predictions. However, each of these methodologies is currently limited in its application. Ensemble modeling is currently limited by computing and data storage resources; data assimilation will require development of solutions suited for the sparse space physics data; and machine learning is currently limited both by data quality and data quantity, particularly concerning extreme events. Machine learning models for space weather produce best results when aimed at both understanding the underlying physical phenomena and producing the more typical machine learning "black box" predictions.
- *Data scarcity* is a real issue: The "too often, too quiet" problem arises because space weather data sets are dominated by quiet conditions, containing only as rare occurrences storms that are of relevance to severe space weather. This creates a serious problem for any machine learning algorithm, and it also poses challenges for defining meaningful metrics that assess the ability of a model to predict interesting but rare events.
- *Interdisciplinary teams, courses, and programs* are needed to train the next generation of space physicists. In particular, it is important to see that a sufficient fraction of the community is equipped with the machine learning skills and knowledge they will need in their research and applied work.

RESEARCH INFRASTRUCTURE

Capabilities have steadily increased in spacecraft development, launch systems, modeling and data analytics, instrumentation, and commercial data product services. This has led to a changed vision for the future science missions' infrastructure and operations: on top of the science priorities to increase understanding of the solar–terrestrial environment, the R2O2R cycles should be included in the planning from the beginning. These emerging trends will play a major role in shaping future architectures. Central themes on the topic of infrastructure are the need for resources and the need to plan ahead.

A key theme that appeared across sessions was the need to increase the operational infrastructure by incorporating data from scientific or (future) commercial missions into operations. This will require, for example, inter-calibration of science instruments so that data sets from different missions can be incorporated into operational models. Preplanning and coordination between the science and operations agencies are needed to address potential issues regarding calibration, data uniformity, and latency.

NOAA has established a Program of Record[5] that identifies the essential observations that must continuously be obtained. These include observations of the Sun, the solar wind, the geostationary environment, and the ionosphere. While academic research can often use data after the fact, some key information about the Sun and the heliosphere needs to be provided in (near) real-time to enable operational decisions.

Commercial data buys have the potential to produce new data products at substantially lower cost and on a faster schedule than existing contractual mechanisms, as launch services, data services, spacecraft systems, and operations are all becoming more commoditized. Moreover, the space weather research community could greatly benefit from gaining access to already existing infrastructure instead of building new infrastructure. An example of such opportunities would be the use of commercial and Department of Defense communication networks to provide data downlink and reduced-latency services.

While commercial data acquisition may offer valuable opportunities, the use of such data faces significant challenges. First and foremost, the vendors must be offered an appealing value proposition to hosting an instrument or providing data, as the task requires additional effort and customization on their part. A further challenge of providing NASA-funded custom instruments to commercial spacecraft is that the U.S. government procurement processes are not well suited to engaging with commercial providers, who tend to move faster and have different commitment processes. Regardless of the instrument developer, there may be proprietary concerns that limit getting the necessary metadata needed to validate and calibrate the data sets. Lastly, after acquisition, resources must be allocated for archiving, validation, verification, and distribution of the data. Cost–benefit analyses will be needed to weigh the issues related to rideshare instrument data quality and the possibly non-optimal distribution of observing locations.

New ways of doing business and managing programs will be needed in order to increase the efficiency of the manufacturing and operation of satellite constellations, as well as the efficiency of their data delivery. Novel approaches to building instruments that can be effectively used for both research and space weather observations should be examined. Systems-level approaches are needed to address gaps arising from the sparse fleet, including coverage across the multiple scales. The large data volumes now being produced require new capabilities for managing, storing, and accessing the data, including techniques for processing data onboard the spacecraft.

Resources and agency coordination will be needed, especially to ensure that instruments will be available and operational as needed. Commercial and other operators will need to be supported and incentivized to support space weather measurements. While cited as not particularly glamorous work, this is necessary to obtain the ground- and space-based infrastructure needed for accurate space weather predictions.

The particular infrastructure needs for space weather include the following:

- *Monitoring the solar source*: Space weather forecasts need more complete (i.e., three-dimensional) coverage of the Sun and heliosphere, along with more comprehensive high-resolution (in both time and energy) measurements of the solar wind. For predicting and understanding the solar corona, it is critically important to continuously monitor the photospheric magnetic field. Such 4π steradian

---

[5] See, for example, National Oceanic & Atmospheric Administration, "Space Weather Program Formulation," https://www.nesdis. noaa.gov/sites/default/files/SessionIII_Talaat_Elsayed_SECOND_SpaceWeatherProgramFormulation_0.pdf, accessed July 8, 2022, or NOAA National Satellite and Information Service, 2021, "NOAA's Current and Future Space Weather Architecture," https://www. nationalacademies.org/event/09-30-2021/docs/DDD4B18BA6DB94017EF5F3F2581987716BB4581A1EFC.

coverage can be obtained by out-of-the-ecliptic architectures that can provide three-dimensional reconstruction of incoming transients. The L4 and L5 vantage points offer unique views, with the L5 monitors more useful for measuring active regions likely to launch coronal mass ejections Earthward, and L4 being more useful for monitoring Earthward-heading solar energetic particles.

- *Observing solar wind heavy ion composition*: Knowledge about the heavy ion composition of the solar wind is a significant gap in current in situ plasma measurements. The composition measurements are key in differentiating high-speed streams from interplanetary coronal mass ejections (ICMEs), in identifying the internal structure of the ICME, and in constraining the models of coronal mass ejection initiation, energization, heating, release, and propagation. They are essential to improve scientific understanding of solar wind structures, which will enable better nowcasting and the quantification of the total mass and pressure of a space weather event.

- *Bridging the gap in heliophysics between the Sun and geospace*: Currently a major gap exists between the study of the Sun and of geospace. To bridge the gap and to predict space weather events from the solar transients to the geospace effects requires new architectures that fuse research in both areas. Such architectures would resolve both the solar source and the solar wind upstream of Earth in scales relevant for the geospace environment.

- *Collecting multipoint observations of the magnetosphere–ionosphere–thermosphere system*: This is a critical need for space weather science and application development, in particular for improving empirical and physics-based models. However, studies of where, what kind, and how many measurements are needed for major advances in the predictions have not been done in a comprehensive way.

- *Transitioning science missions to operational ones*: The completion of the prime science phase of a science mission offers opportunities to continue monitoring the Sun–Earth system with additional observation points. Designing science missions from the outset with both scientific and operational targets in mind would ease the transition process. For example, the CCOR (Compact Coronagraph) was developed in a collaboration between the Naval Research Laboratory and NOAA to be flown on both the Space Weather Follow On-Lagrange 1 (SWFO-L1) mission and GOES-R at geosynchronous orbit. This interagency collaboration is intended to continue a critical space weather measurement series initiated by the ESA/NASA Solar and Heliospheric Observatory (SOHO) white light coronagraph launched in 1995. The project is a model for interagency coordination as well as instrument development and deployment within agency budgets.

1

The Space Weather Community

Setting the stage for subsequent sessions, the workshop began with an overview of the space weather enterprise, with particular attention given to the drivers of a rapidly changing landscape of space weather activities throughout the government (Box 1-1). Of particular importance was the impact of the passage of the Promoting Research and Observations of Space Weather to Improve the Forecasting of Tomorrow (PROSWIFT) Act in December 2020.[1]

The session began with a keynote address that summarized the proceedings of the Phase I Space Weather Workshop held in 2020 (NASEM 2021). The following four panels each focused on a different aspect of the space weather community: major government agencies involved in space weather research and operations, details about interagency partnerships and collaboration, the user community and operations, and the space weather workforce, with a particular emphasis on workforce diversity. This chapter summarizes the presentations and discussions in each of those four panels.

---

[1] Among its key provisions, is the PROSWIFT Act's delineation of federal agency roles and responsibilities for space weather: The "transition to operations" is addressed by the PROSWIFT Act in part through the creation of the Space Weather Interagency Working Group (SWAG). The SWAG is tasked with developing formal mechanisms to transition space weather research findings, models, and capabilities of NASA, NSF, and the U.S. Geological Survey, and other relevant federal agencies as appropriate, to NOAA and the Department of Defense. The Act also creates a "Government-University-Commercial Roundtable on Space Weather," hosted by the National Academies, to facilitate communication and knowledge transfer among government participants in the Space Weather Operations, Research, and Mitigation (SWORM) Interagency Working Group, the academic community, and the commercial space weather sector. The PROSWIFT Act also directs NOAA to develop "near real-time coronal mass ejection imagery, solar wind, solar imaging, coronal imagery, and other relevant observations required to provide space weather forecasts." In addition, it directs NOAA to consider enhancement of its space weather capabilities via "commercial solutions, prize authority, academic partnerships, microsatellites, ground-based instruments, and opportunities to deploy the instrument or instruments as a secondary payload on an upcoming planned launches." See, U.S. Congress, 2020, PROSWIFT Act: Promoting Research and Observations of Space Weather to Improve the Forecasting of Tomorrow Act, S.881 – 116th Congress (2019-2020), Public Law 116-181, https://www.congress.gov/bill/116th-congress/senate-bill/881.

BOX 1-1
Recent Agency Actions and Activities to Support Space Weather

NASA Actions
- Space Weather Science Application Program (SWxSA) is being set up within the Heliophysics Division (HPD).
- HPD commissioned a space weather science and measurement gap analysis that was performed by a committee of space weather experts from academia, the commercial sector, and the space weather operational and end-user community under a NASA task order to the Johns Hopkins University Applied Physics Laboratory.
- The first space weather "proving ground" (the Architecture for Collaborative Evaluation) was established under the Space Weather Research-to-Operations and Operations-to-Research (R2O2R) Framework[a] and is hosted by NASA Goddard Space Flight Center.

NOAA Actions
- Space Weather Prediction Testbed (SWPT) is planned to be located at NOAA's Space Weather Prediction Center to accelerate the improvement of operational space weather services under the R2O2R Framework.[b]
- A coupled ionosphere/thermosphere model has been introduced into operations at the NOAA Space Weather Prediction Center (SWPC).
- An Open-Architecture Data Repository (OADR) prototype environment is developed by NOAA Office of Space Commerce to facilitate R2O transition related to space traffic management. The OADR is a partnership between industry, government, and academia.
- Operational coronal images will be provided by NOAA's Space Weather Follow On-Lagrange 1 (SWFO-L1) mission, whose launch is planned for 2025.

NSF Actions
- A research opportunity "Advancing National Space Weather Expertise and Research toward Societal Resiliency (ANSWERS)" was established.
- A focus area on data infrastructure needs is being developed.

Interagency Actions
- NASA and NOAA signed an interagency agreement to provide space radiation environment support for the conduct of all human spaceflight.
- NOAA and NASA, with SWORM (Space Weather Operations, Research, and Mitigation subcommittee) support, have formalized an R2O2R Framework.
- As directed in the PROSWIFT (Promoting Research and Observations of Space Weather to Improve the Forecasting of Tomorrow) Act, the Space Weather Advisory Group is developing a user needs survey for space weather products.

---

[a] Executive Office of the President, 2022, "Space Weather Research-to-Operations and Operation-to-Research Framework," Space Weather Operations, Research, and Mitigation Subcommittee Committee on Homeland and National Security of the National Science and Technology Council, https://www.whitehouse.gov/wp-content/uploads/2022/03/03-2022-Space-Weather-R2O2R-Framework.pdf.
[b] The President's FY 2022 budget requested $5 million for the SWPT, with construction to begin in 2022. However, the enacted budget for NOAA did not include SWPT funding.

AGENCY UPDATES

The Agency Panel provided updates (since the Phase I proceedings) on the priorities and activities of the various government agencies involved in space weather and on the coordination among them. The panelists were Jim Spann, the space weather lead in NASA's Heliophysics Division; Mangala Sharma, program director for space weather research in the Directorate of Geosciences Division of Atmospheric and Geospace Sciences (GEO/AGS) at the National Science Foundation (NSF); and Elsayed Talaat, Director of the Office of Projects, Planning, and Analysis at the National Environmental Satellite, Data, and Information

Service (NESDIS) of the National Oceanic and Atmospheric Administration (NOAA). They were asked to address the following four key questions:

- How will we take advantage of the recently established interagency focus on space weather, and what new interfaces may be needed for space weather research and fluent (i.e., smooth and effective) service structure?
- What are the new communities that we need to engage in space weather, and what are the mechanisms to involve them?
- What should the education of next-generation space weather scientists and forecasters look like?
- What should the engineers in different fields know about space weather?

National Aeronautics and Space Administration

In the first presentation, Spann described NASA's role in space weather. He noted that NASA is setting up a dedicated Space Weather Program within its Heliophysics Division, which will modify how the agency internally funds space weather activities. Spann described NASA's role in the space weather enterprise as advancing research by providing unique, significant, and exploratory data streams for research building on theory, modeling, and data analysis as well as for operations making use of the data. The agency's programs are designed to explore observing techniques, models, and data analysis so that once the potential value of a new capability to space weather services has been demonstrated, that capability can become a candidate for transition to operations. Thus, he said, NASA and its Heliophysics Division is "uniquely poised to support the needs of the national and international space weather enterprise." The new NASA Space Weather Program is intended as a "national resource to unify space weather research and drive our understanding of its risks, impacts, and mechanisms."

NASA has established four pillars (Figure 1-1) that support space weather within the Heliophysics Division: Investigation, Transition, Exploration, and Application. Each of these pillars has a particular

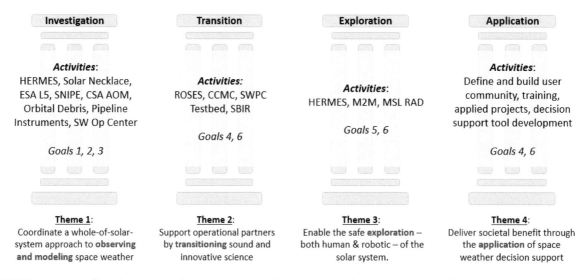

FIGURE 1-1 Four pillars that support the new Space Weather program within the NASA Heliophysics Division. The six goals of NASA's Space Weather Strategy can be found at https://science.nasa.gov/heliophysics/space-weather-strategy.
NOTE: Acronyms defined in Appendix D.
SOURCE: Jim Spann, NASA, presentation to workshop, April 11, 2022.

theme describing its goals. The Investigation pillar represents the fundamental scientific investigations that NASA supports with the goal of providing a coordinated solar system research approach to observing and modeling space weather. The Transition pillar supports operational partners by transitioning the innovative science developed through research into operational (space weather) capabilities. The Exploration pillar involves safe exploration of the solar system, through both human and robotic means, including plans for the return of crewed missions to the Moon and subsequently to Mars. The Heliophysics Division's role in Exploration is to provide support through understanding of the space environment. Finally, the new, less mature, Applications pillar supports new tools with a long-term aim to help mitigate space weather impacts.

The NASA Space Weather Program has a number of current and planned activities that will contribute to the four pillars. For example, in support of Exploration, the HERMES (Heliophysics Environmental and Radiation Measurement Experiment Suite) instrument package to be placed on the lunar-orbiting Gateway will make space weather measurements in support of lunar operations, and demonstrate technologies needed to conduct human missions to Mars. To support the Investigation and Transition pillars, space weather R2O2R (research-to-operations/operations-to-research) grants, space weather centers of excellence, and Small Business Innovative Research opportunities will fund targeted research to develop applications and respond to operational needs. The Heliophysics System Observatory observations made for basic and applications-focused research can, in some cases, provide real-time support for operational space weather applications. NASA also funded the recently completed Space Weather Gap Analysis (NASA 2021), which identified high-priority observations that are at risk of being decommissioned, not currently available, or that are needed to significantly advance space weather forecasting and nowcasting capabilities.

National Science Foundation

Sharma stated that, consistent with the broader NSF missions, the agency's space weather agenda is focused on fundamental science with broader impacts, on the infrastructure needed to achieve the scientific goals, and on the people engaged in the science. Furthermore, the space weather–related activities contribute to NSF's goals to empower science, technology, engineering, and mathematics (STEM) talent to fully participate in science and engineering and to create new knowledge and to translate that knowledge into solutions for the benefit of society.

Space weather research at NSF is supported by multiple programs. Within the Directorate of Geosciences, the Division of Atmospheric and Geospace Sciences supports the geospace research programs (e.g., Coupling, Energetics, and Dynamics of Atmospheric Regions [CEDAR]; Geospace Environment Modeling [GEM]; and Solar, Heliospheric, and Interplanetary Environment [SHINE]) and facilities such as the National Center for Atmospheric Research (NCAR). Within the Directorate for Mathematical and Physical Sciences, the Division of Astronomical Sciences supports astronomy and astrophysical research, and the National Solar Observatory facility. The Division of Physics supports plasma physics research relevant to Heliophysics.

Recognizing space weather as a grand challenge in geosciences, Sharma described NSF's recently launched research opportunity, Advancing National Space Weather Expertise and Research toward Societal Resiliency (ANSWERS). This program seeks to develop a deep and transformative understanding of the dynamic, integrated Sun–Earth system and the solar and terrestrial drivers of space weather, as well as to link geospace observers, theorists, modelers, software developers, laboratory experimenters, STEM educators, and space weather policy experts together.

Development of the space science workforce is an important priority to NSF. To that end, NSF has made a number of early-career faculty development awards covering a wide range of disciplines relevant to space weather, from the dynamics of the solar corona to satellite debris in low Earth orbit.

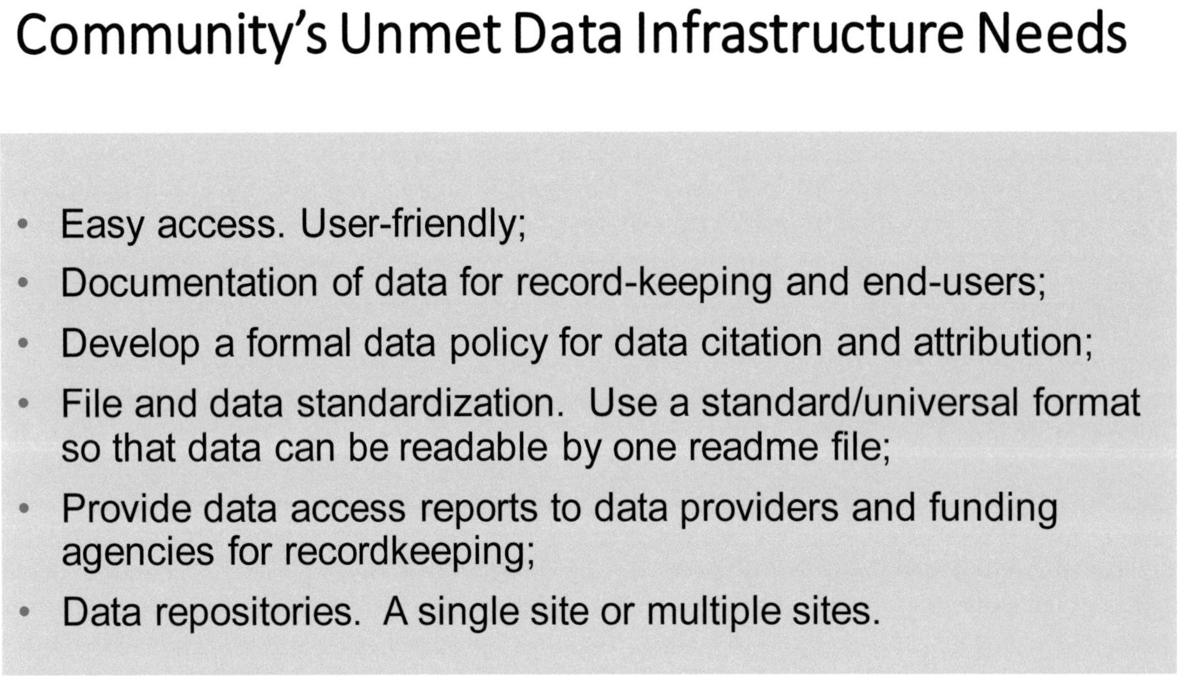

Community's Unmet Data Infrastructure Needs

- Easy access. User-friendly;
- Documentation of data for record-keeping and end-users;
- Develop a formal data policy for data citation and attribution;
- File and data standardization. Use a standard/universal format so that data can be readable by one readme file;
- Provide data access reports to data providers and funding agencies for recordkeeping;
- Data repositories. A single site or multiple sites.

FIGURE 1-2 The National Science Foundation's new focus on data infrastructure, laying out the needs of the space weather community as it relates to data curation and availability.
SOURCE: Carrie Black, National Science Foundation, presentation to workshop, April 13, 2022.

NSF supports a wide network of observing facilities, which cover the full breadth of the Sun–Earth domain. The facility suite includes solar telescopes, magnetometer networks, coherent and incoherent radars, a neutron monitor network, and networks of citizen-science measurements. Although the NSF facilities are primarily ground-based, NSF also supports balloon-based instrumentation, CubeSats, and public–private partnerships within the space industry, such as the Active Magnetosphere and Planetary Electrodynamics Response Experiment (AMPERE). NSF is also starting to focus more heavily on data infrastructure needs (Figure 1-2).

National Oceanic and Atmospheric Administration

NOAA's space weather activities as described by Talaat are based on the agency's charter, which calls for NOAA to build capacity to advance space weather policy, to accelerate growth in NOAA space weather services, and to apply an integrative and collaborative approach between space weather research and operations. This charter is complementary to those of NASA and the NSF. As mentioned above, space weather policy activities include the implementation of the National Space Weather Strategy and Action Plan and responding to the actions detailed in the PROSWIFT Act. Improving space weather services involves sustaining fundamental observations, providing accurate models and forecast products, transitioning scientific and technological advances into operations, and supporting the private sector to fill data and technology gaps and to provide value-added services.

In order to improve services and address user needs, NOAA advances a space weather paradigm that is parallel to the terrestrial weather one, Talaat said. The previous paradigm relied heavily on research observations, lacked a formal framework for community input on R2O2R efforts, and emphasized global geomagnetic activity indices as indicators of space weather state. The new paradigm combines the establishment of dedicated operational observing systems with modeling and service improvements to provide the forecasts, warnings, and data necessary to protect the nation's critical infrastructure. The formal R2O2R Framework described in the next section will incorporate contributions from industry, government agencies, and academic partners to test and evaluate emerging science, accelerate the transition of new capabilities to operations, and enable the improvement and maintenance of existing operational models. Regional and local specifications and forecasts tailored to the needs of decision-makers within the industries will augment global indices.

NOAA currently uses a suite of operational numerical models covering the Sun–Earth system, and these capabilities will be further enhanced through R2O2R efforts. Models predicting the solar wind and the propagation of coronal mass ejections have been operational since 2011 and were upgraded in 2019. The Space Weather Modeling Framework (SWMF) Geospace model used to predict regional geomagnetic activity has been operational since 2016 and was upgraded in 2021. A model for the regional electric fields and associated currents within electric power grids has been operational since 2020, and the coupled Whole Atmosphere Model–Ionosphere Plasmasphere Exosphere (WAM-IPE) model became operational in 2021. Operational models now also provide advisories for civil aviation, including a human radiation dose specification model. In addition, NASA and NOAA recently signed an interagency agreement on NOAA providing space radiation environment support for all human spaceflight activities. This continues NOAA's long-time support for NASA's crewed space activities, including 24/7 space weather forecasts and alerts for the International Space Station, Artemis, Lunar missions, Lunar surface operations, and future Mars missions.

The goal of NOAA's National Environmental Satellite, Data, and Information Service (NESDIS) is to provide an integrated, digital understanding of Earth's environment in a way that can evolve rapidly to address changing user needs. This is accomplished by combining NOAA's assets with those of partners. Recently, space weather has been elevated to a third pillar in NOAA's observing infrastructure, along with geostationary and low Earth orbit systems, and NESDIS is working to integrate capabilities with its national and international partners. Space weather conditions will be monitored at the L1 Lagrange point, geostationary orbit, and low Earth orbit, and, potentially, at the L5 Lagrange point in the future. A set of common ground data services will verify, calibrate, and fuse data into improved products and services.

NOAA has established a program of record that identifies the essential observations that must continuously be obtained (see Figure 1-3). These include observations of the Sun, the solar wind, the geostationary environment, and the ionosphere, which are all in operational use today. An essential mission for maintaining the continuity of solar and solar wind measurements is the Space Weather Follow On-Lagrange 1 (SWFO-L1). This mission is currently in development, and the spacecraft and instrument critical design reviews were recently completed. The SWFO-L1, whose launch is planned for February 2025, will carry a compact coronagraph to provide operational coronal images that are now obtained from the Solar and Heliospheric Observatory (SOHO) research mission launched in 1995. The compact coronagraph will also be included on the Geostationary Operational Environmental Satellite (GOES-U) spacecraft to be launched in 2024. In addition, GOES-T, which was successfully launched on March 1, 2022 (renamed GOES-18 upon becoming operational), carries the standard set of space weather instruments similar to those onboard GOES-16 and -17, and will replace GOES-17 in January 2023. Another recent achievement was to reduce the latency of ionospheric data obtained from the Constellation Observing System for Meteorology, Ionosphere, and Climate (COSMIC-2) mission to under 30 minutes (Weiss et al. 2022). These measurements of the ionospheric total electron content have been acquired since 2019, and are being used in NOAA's operational model for the ionosphere.

NOAA/NESDIS Formulating a Space Weather Program

- Diverse observing requirements must be made from diverse vantage points (LEO, GEO, Sun-Earth line, L1 and off the Sun-Earth line)
- Continuity and anticipated product improvement need dates are varied:
  - ○ Long Lead Instrumentation
  - ○ Next Generation L1 & off-Sun-Earth-axis
  - ○ Space Weather Ground Operations
  - ○ Geostationary Observations
  - ○ Tundra/High Elliptical Orbit Observations
  - ○ Low Earth Orbit Observations
- Program formulation will initialize a loosely coupled program with an initial set of projects.

FIGURE 1-3 National Oceanic and Atmospheric Administration (NOAA) program to determine necessary observations for their operational space weather needs.
NOTE: GEO = geosynchronous orbit; LEO = low Earth orbit; NESDIS = National Environmental Satellite, Data, and Information Service.
SOURCE: Elsayed Talaat, National Oceanic and Atmospheric Administration, presentation to workshop, April 11, 2022.

R2O2R support at NOAA includes multi-agency coordination with its Space Weather Prediction Testbed and the interagency Community Coordinated Modeling Center (CCMC). A major upcoming exercise will be focused on understanding the needs of the aviation community and identifying steps for the improvement of targeted operational services.

INTERAGENCY PARTNERSHIPS: NEW WAYS OF WORKING

The Interagency Partnerships Panel examined the issues that arise when multiple organizations are responsible for space weather and, in particular, ways to improve coordination and communication across these agencies. The four panelists were Jinni Meehan, a program manager of NOAA's National Weather Service (NWS); Dan Moses, a program scientist in NASA's Heliophysics Division; Tammy Dickinson, the president of Science Matters Inc.; and Sage Andorka, a lead systems engineer for the U.S. Space Force. The panel was asked to address the same four key questions as the previous panel.

The panelists provided updates on the different national programs involving space weather and on the current priorities for addressing space weather challenges. As Meehan explained, the central strategy of the U.S. national space weather effort (NSW-SAP) was put in place in 2019, thereby superseding the 2015 strategy and action plan documents. Space weather activities are overseen by the Space Weather Operations, Research, and Mitigation (SWORM) Subcommittee, an interagency working group that is organized under the National Science and Technology Council (NSTC) Committee on Homeland and National Security, under the Office of Science and Technology Policy (OSTP).

The PROSWIFT Act signed on October 21, 2020, codified SWORM into law. It also codified the policy of the United States to prepare and protect against the societal and economic impacts of space weather. It established an interagency working group (IWG) to coordinate executive branch actions to improve

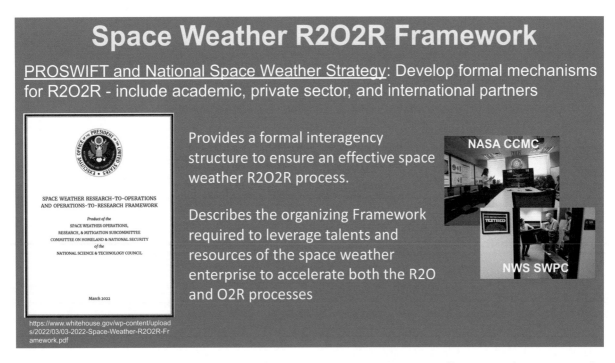

FIGURE 1-4 Highlights of the Space Weather Operations, Research, and Mitigation's effort to provide structure to the R2O2R framework.
NOTE: CCMC = Community Coordinated Modeling Center; NWS = National Weather Service; PROSWIFT = Promoting Research and Observations of Space Weather to Improve the Forecasting of Tomorrow Act; R2O2R = research to operations/operations to research; SWPC = Space Weather Prediction Center.
SOURCE: Jinni Meehan, NOAA/National Weather Service, presentation to workshop, April 11, 2022.

understanding and prediction of space weather phenomena, and it also established a Space Weather Advisory Group (SWAG) (NOAA 2021) to receive advice from academia, the commercial sector, and space weather end users. Meehan noted that the PROSWIFT Act does not authorize agency funding, but rather serves as a means to prioritize agency efforts.

In March 2022, just a few weeks before the workshop, OSTP released the Space Weather Research-to-Operations and Operations-to-Research Framework (Executive Office of the President 2022) to the public. The framework was developed in response to one of the National Space Weather actions, and it provides a formal interagency structure aimed at ensuring an effective space weather R2O2R process. The SWORM Subcommittee is now in the process of updating the 2019 National Space Weather Strategy and Action Plan with priorities including the R2O2R Framework (Figure 1-4), benchmarks, activity scales, hazard mapping, human exploration and aviation fields, human resources through the Solar System Ambassador program, and the continuity of availability of satellite observations. However, many of these plans still lack concrete implementation strategies.

Moses reviewed NASA's role in space weather, noting the various executive (e.g., NSW-SAP) and legislative (e.g., the PROSWIFT Act) mandates that direct the agency to address space weather research and applications. NASA's Heliophysics Division works as the research arm of the U.S. space weather effort, he said, and, in particular, it coordinates the National Space Weather Action Plan with various other agencies, including NOAA, NSF, the U.S. Geological Survey, and the U.S. Air Force Research Laboratory.

Dickinson introduced the Space Weather Advisory Group (SWAG) put in place by the PROSWIFT Act. The SWAG, chaired by Dickinson, has 15 members appointed by SWORM, five each from academia, commercial space weather, and end-user communities. The group is an important new asset for the SWORM Subcommittee, she said, as it gets more direct advice from the commercial sector on space weather priorities. The SWAG's first task, which is now being addressed, is to conduct a user needs survey concerning space weather products. This survey will address 10 different sectors (the electric power grid, satellites, aviation, human space flight, etc.), and develop sector-specific surveys for each of them.

In addition to the SWAG, two other space weather–related groups have been recently formed: the Space Weather Roundtable[2] of the National Academies of Sciences, Engineering, and Medicine (the National Academies), which was also required by the PROSWIFT Act, and the NASA Space Weather Council, which is a subgroup of the Heliophysics Advisory Committee. These two entities and the SWAG will support communication and coordination of space weather efforts among the agencies.

Andorka offered a broad overview of the U.S. Space Force interests in space weather, including opportunities for collaboration. The Department of Defense space weather program has recently been reorganized within the U.S. Space Force under Space Domain Awareness. Because of its vital importance for space warfighters, the Space Force is developing the Space Domain Awareness Environmental Toolkit for Defense (SET4D; see Andorka et al. 2021), which will use space weather information together with information about space warfighter missions. The goal of this effort is to create a holistic picture of the operational environment, including specific system impacts. SET4D development will require an integrated software suite consisting of data, models, and applications.

In support of the space domain awareness efforts, the Space Force is establishing a number of independent interagency agreements, which will cover data sharing between agencies, synchronizing data stores, establishing and using open architectures, and enabling all agencies to participate in capability improvements. An important interagency collaboration under discussion is the one involving NASA and NOAA on commercial space traffic management. Collaborations are also being pursued between commercial entities and academia to feed operational needs into the research environment (O2R).

USER COMMUNITY AND OPERATIONS

The next panel covered the space weather user community and operations. The five panelists were Hazel Bain, a research scientist at NOAA's Space Weather Prediction Center (SWPC) and the University of Colorado; Michele Cash, a research section lead at SWPC; Mark Olson, a senior engineer and manager at the North American Electric Reliability Corporation; Michael Stills, a former director of flight dispatch at United Airlines; and Scott Leonard, the technical director at NOAA's Office of Space Commerce. The panelists were asked to address three key questions:

- What new observations, models, or other assets are needed to supply high-quality services?
- How do we educate the user community to be able to demand new services making use of novel observations and models?
- Where are the competence gaps in the community?

The speakers focused on the much-needed ongoing efforts to transition research capabilities to operations, identifying a subset of key user needs. The government agencies' commitment to improving operational services will be accomplished by developing and transitioning new or updated space weather capabilities in partnership between academia and commercial enterprises. As described in the R2O2R Framework document (see the

---

[2] See National Academies of Sciences, Engineering, and Medicine, 2022, "Space Weather Roundtable," https://www.national academies.org/our-work/space-weather-roundtable.

previous subsection), operational services will be improved by an efficient transitioning of capabilities from research to operations; by engagement of forecasters and other operators in evaluation, testing, and feedback; and by using specific operational experiments to determine the efficiency and impact of the new capabilities.

Bain said that an important component in improving current operational models is to establish and communicate their current baseline capabilities. Such benchmarking will enable the research community to assess the merits of new capabilities as well as enable the operational agencies to prioritize transition activities. Examples include the recent validation of solar proton forecasts, and quantification of the impact of spatially distributed data in an ionospheric disturbance model.

In addition to transitioning research algorithms and models into operations, Bain also discussed the need to use research observations in operations. As the operational models rely on increasing amounts of research data, their availability for operations must be ensured. For example, the research-focused global neutron monitoring network that is used as an input to operational radiation models for the aviation industry has no guarantee for long-term continuity of its operations.

Cash expanded on Bain's presentation, providing detail on the research-to-operations (R2O) process and on what is needed to transform research results into reliable operations. Detailing a formal R2O approach, she described a series of readiness levels (RLs) used to indicate the maturity of a given capability, and the process to advance from lower readiness levels to higher, more operational, readiness levels (Figure 1-5). The first space weather proving ground being established under the R2O2R Framework is the Architecture for Collaborative Evaluation hosted at NASA's Goddard Space Flight Center. The first space weather testbed is being established at NOAA's SWPC.

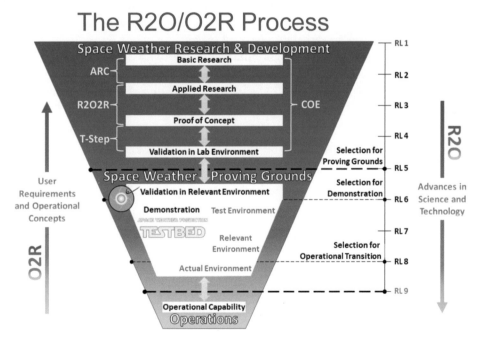

FIGURE 1-5 The space weather R2O2R funnel, showing the processes by which advances in research transition to operational capabilities.
NOTE: ARC = Applied Research Challenge; COE = Center of Excellence; R2O2R = research to operations/operations to research; RL = readiness level.
SOURCE: Michelle Cash, NOAA Space Weather Prediction Center, presentation to workshop, April 11, 2022.

NOAA has a long history of using testbed experiments to facilitate the transition of research capabilities into operations, and the first space weather testbed will conduct its initial experiment in September 2022. These experiments bring together forecasters, users, regulators, internal and external researchers, and federal partners to explore current capabilities, needs, and gaps in the existing space weather services. Such experiments support emerging concepts, demonstrate new technologies, and provide feedback to developers.

Although not all space weather user needs could be comprehensively discussed in the workshop, representatives from the electric power and aviation industries provided perspectives on how space weather information is used and the types of information that can improve the resiliency of these industries. The industry representatives said that the collaborative engagement between industry, researchers, and service providers has been essential for the progress.

Olson began with a brief description of the North American Electric Reliability Corporation (NERC), describing its tasks of developing and enforcing standards for the North American electrical grid and minimizing the likelihood of grid failures and blackouts. In particular, two mandatory reliability standards have been implemented to reduce the risks associated with geomagnetic disturbances (GMDs): First, grid operators must have procedures to mitigate impacts during GMD events, and second, grid planners and asset owners must assess and design the system to mitigate a rare, high-intensity event with probability of occurrence once in 100 years.

The electric power industry has identified a number of research efforts that would make space weather information more actionable. One key need is to have more granularity to the storm intensity scales. The highest level of geomagnetic activity, characterized by a NOAA G-scale of 5, covers a broad range of disturbance magnitudes. The smaller G5 storms may have only minor impacts on the electrical grid, whereas strong G5 storms may cause major impacts. However, as these events are so rare, developing such scales will need to rely on more research to develop models that can cover such conditions. Further needs include longer forecast lead times, more accurate determination of the geographic location and spatial extent of the events, and better uncertainty and confidence estimates.

Improving the electric power grid design and architecture and addressing the vulnerabilities of the current grid requires knowledge of the details of the space weather disturbance (peak magnitude and location of severe events), of the ground geoelectric properties (the geoelectric field spatial and temporal granularity, the 3D structure of the ground conductivity), as well as further research to get improved estimates of induced currents in regions with complex conductivity distributions. NERC has initiated a data collection program to support geomagnetically induced current (GIC) model validation. Data on induced currents and magnetic field variability have been collected for strong events since May 2013 and will be made available through a public portal beginning in mid-2022.

The North American electric industry recognizes the benefits from a strong collaboration between the space weather community and industrial stakeholders, which has facilitated both R2O and O2R processes. This collaboration has been facilitated by regular information-exchange meetings involving industry representatives, researchers, and service providers.

Stills said that the aviation industry's major focus is on polar routes where U.S. (and other) carriers are required to monitor space weather and to have a plan for increased energetic particle doses. For example, United Airlines has flown polar routes since the year 2000 and has had numerous occasions when space weather mitigation has altered flight routes or altitudes. The airline industry's desire to directly relate the environmental conditions along a route to the doses affecting the aircraft and its passengers could be achieved by collecting data during flights and making them available for model validation (see a later discussion on collection and sharing of data from commercial providers). Better data and models to support decision making would avoid flight planners' overreaction to space weather warnings. As is the case for most space weather users, forecast lead time is important for the airline industry: A 72-hour lead time would allow the airlines to reschedule both passengers and cargo. However, shorter lead-time warnings are also useful, as they indicate when communication integrity may be affected. Since U.S. carriers are required to always maintain communication with aircraft, potentially affected aircraft need to get warnings about space weather communication hazards in a timely manner.

Stills stressed education as an important element in improving the aviation industry's response to space weather. Both Airlines for America and NOAA have been involved in providing baseline education to various carriers. Expanding such efforts to air navigation services providers, including oceanic air traffic controllers, could serve to establish common protocols for space weather events. In addition, a standard knowledge package should be distributed throughout the airline industry to ensure that consistent procedures are followed by all carriers.

Leonard described the Open-Architecture Data Repository (OADR), a prototype environment for R2O established at NOAA's Office of Space Commerce (Figure 1-6). The OADR was developed in partnership with industry, government, and academia in response to Space Policy Directive-3, which includes the mandate to build an operational system to provide basic space situational awareness (SSA, a subset of space domain awareness) information, primarily for collision avoidance and space traffic management of satellites in low Earth orbit. The proving ground will be used to expand and mature SSA stakeholder and system requirements, support ongoing research and experimentation, and serve as a development environment for further operational applications and algorithms.

The OADR architecture is cloud based and scalable to accommodate the increased number of satellites being launched in the near future, Leonard said. The modular design will be adaptable to support new technologies, advanced algorithms, and diverse data sets, including space weather measurements as well as satellite orbit information. A priority is to promote data sharing and collaboration across the U.S. and international space weather communities.

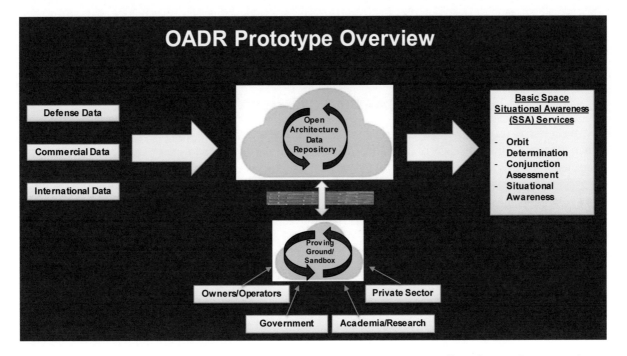

FIGURE 1-6 Overview of the National Oceanic and Atmospheric Administration's Office of Space Commerce Open-Architecture Data Repository (OADR).
SOURCE: Scott Leonard, NOAA Office of Space Commerce, presentation to workshop, April 11, 2022.

**Demographics
Space Sciences
Workforce**

• Women slowly
  catching up

• Non-White racial
  groups remain
  severely under-
  represented

• Why so low?
  Why so slow?

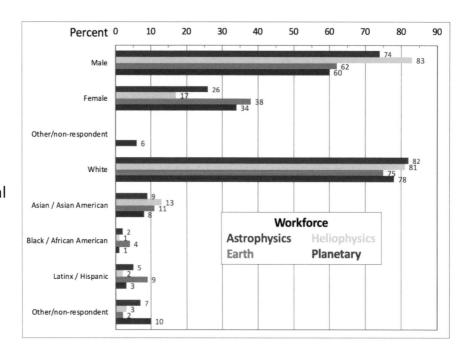

FIGURE 1-7 Demographic representation in the space science workforce. The demographics are broken down by four of the divisions within NASA Science Mission Directorate: Astrophysics, Heliophysics, Earth Science, and Planetary Science. SOURCES: Fran Bagenal, University of Colorado Boulder, presentation to workshop, April 11, 2022; from National Academies of Sciences, Engineering, and Medicine, 2022, *Advancing Diversity, Equity, Inclusion, and Accessibility in the Leadership of Competed Space Missions*, Washington, DC: The National Academies Press, https://doi.org/10.17226/26385.

DIVERSITY IN WORKFORCE

The session's final panel dealt with diversity in the space sciences workforce. The three panelists were Frances Bagenal of the University of Colorado, MacArthur Jones of the Naval Research Laboratory, and Edward Gonzalez of NASA's Goddard Space Flight Center. The panelists were asked to address three key questions:

• What is the current state of the demographics within the space sciences workforce?
• What are the reasons for the lack of diversity in the geosciences?
• When we have a diverse workforce, how do we retain it?

In response, the panel discussed various problems and potential solutions related to equity and diversity.

Bagenal began by offering details on the demographics of the physics workforce (Figure 1-7). A substantial amount of information has been obtained through workforce surveys compiled by the statistical division of the American Institute of Physics (AIP).[3] Some of the surveys have been done in association with recent decadal surveys, including the astrophysics and planetary decadal surveys. Information has also been obtained from surveys conducted by the National Science Foundation (NSF 2017) and surveys using email lists from the American Geophysical Union, the American Astronomical Society, and the Space Weather Workshop (Moldwin and Morrow 2021).

---

[3] See American Institute of Physics, 2022, "Statistical Research Center," https://www.aip.org/statistics.

The dominant situation shown in all space science disciplines is that the workforce is predominantly white and male. Female representation has been slowly catching up, but non-white racial groups remain severely underrepresented.

Undergraduate education is a major barrier to careers in STEM, Bagenal said. It has been known for decades that a large number of students who begin taking math and physics in a university will eventually drop that major, with only about 20 percent of all math and physics undergraduate majors persisting to graduation, and this is particularly prevalent among underrepresented populations (Bradforth et al. 2015; Seymour and Hunter 2019). First-year mathematics and physics courses are especially important for retaining students in the sciences.

On the topic of doctoral degrees in physics, Bagenal emphasized two facts that have persisted over several decades: First, the number of Latino/Hispanic and African American Ph.D.s is tiny,[4] and, second, the numbers of U.S. and non-U.S. physics Ph.D.s are roughly equal. Even if the global talent pool is large and attractive to U.S. universities, Bagenal questioned the strategy of so strongly relying on foreign students to provide the United States with a highly educated workforce. She urged the United States to improve its domestic pipeline to prepare for a possible decrease in the number of foreign students coming to the United States.

A similar demographic situation is seen in surveys of the scientists submitting research proposals to NASA collected through online personal profiles solicited at the time of proposal submission (Barbier and Wilson 2021). Although the data are somewhat limited, the share of women submitting proposals is ~20 percent, and the Latino/Hispanic and African American populations are severely underrepresented. The numbers are similar across all divisions within NASA's Science Mission Directorate.

Bagenal made one recommendation to expand and improve the demographics data: it is important, she said, that the upcoming Heliophysics Decadal Survey include a panel on the state of the profession. Along these lines, the recent Decadal Survey for Astronomy and Astrophysics stressed that collecting, evaluating, and acting on demographic data was one of its seven essential goals. The 2011 AIP Solar, Space, and Upper Atmospheric Physicists Survey should be used as a benchmark point, and Bagenal suggested that the Heliophysics Decadal Survey should consider employing the AIP to conduct a demographics survey. Data from proposals submitted to NASA need to improve, and these data should be combined with similar data from NSF.[5]

Jones pointed out that it is also essential that there is consistency among the data being collected by various organizations. For example, demographic information recently collected by the CEDAR community was compiled for various career stages: undergraduate, graduate, early-career, mid-career, and senior-career, while some other NSF surveys have focused solely on one group (e.g., postdoctoral researchers), making apples-to-apples comparisons difficult.

Jones spoke about the Significant Opportunities for Atmospheric Research and Science (SOARS) undergraduate-to-graduate bridge program, which strives to get historically underrepresented groups involved in geosciences, especially atmospheric sciences. This program, centered at the NCAR, recognizes the many reasons for the current lack of diversity: Lack of awareness of career opportunities; lack of mathematics and science preparation in underserved secondary schools; feelings of isolation, stereotype threat, and imposter syndrome; implicit biases, including those in recommendation letters; lack of representation and a sense of belonging; and the accumulation of many disadvantages, including systemic racism.

Applicants to the SOARS program reflect some of the situations that can be responsible for the low participation rate of the historically underrepresented groups in the atmospheric and space sciences: Applicants may have majors that are not fully aligned with traditional atmospheric science degrees, or they may have changes or interruptions during their academic pathways. Some have lower GPAs, especially in their first year of college, and they may have variable—and perhaps biased—letters of recommendation.

---

[4] Ibid.
[5] See also NASEM (2022).

Some of these students might be working while in school to support themselves or their families. Thus, asking applicants about their situations can help mentors to understand their unique situation and potential as well as to better target actions to address underrepresentation.

Expanding the recruitment base to minority-serving institutions, tribal colleges, historically Black colleges and universities, and various national conferences is also key to mitigating underrepresentation. Furthermore, it is important to encourage a wider population to take science courses, which can help diversify the workforce at all career paths within and outside academia.

Retention is a key consideration. Job applicants, students, and employees desire an inclusive culture. Humans have a need for a sense of belonging created by seeing people like them throughout the organization; they want to feel included and listened to; and they want to have active employee resource groups, affinity groups, and effective mentoring programs available to them. Such groups can serve as forums to raise the issues that are important to everyone.

As examples of effective retention programs, Gonzalez mentioned the employee resource and affinity groups at NASA Goddard Space Flight Center. Numerous resource groups are active among the Latino/Hispanic employees, Native Americans, African Americans, members of the LGBTQ+ community, Asians, veterans, and so forth. Each group includes champions at the leadership level who attend the meetings, hear the issues and suggestions, and then act on them.

Mentorships and apprenticeships are also effective means to strengthen diversity, Gonzalez said, adding that NASA Goddard offers a good example by providing strong mentoring at multiple levels within the organization. Beyond NASA, the American Geophysical Union also has a mentoring program called 360. Often the most effective approach is what might be called "apprenticeships," which involve long-term, multi-year partnerships (e.g., taking students from high school through college and graduate school and into their professions).

REFERENCES

Andorka, S.K., J.M. Cox, and M.F. Fraizer. 2021. "SDA Environment Toolkit for Defense (SET4D) – Enabling Attribution for Orbital Assets and Electro-Magnetic Spectrum Links Through Streamlined R2O." Advanced Maui Optical and Space Surveillance Technologies Conference (AMOS). https://amostech.com/TechnicalPapers/2021/Poster/Andorka.pdf.

Barbier, L., and C. Wilson. 2021. "Summary Demographic Data." Presentation to the Committee on Increasing Diversity and Inclusion in the Leadership of Competed Missions. June 16. Washington, DC: National Academies of Sciences, Engineering, and Medicine. https://www.nationalacademies.org/event/06-16-2021/increasing-diversity-and-inclusion-in-the-leadership-of-competed-space-missions-meeting-6.

Bradforth, S., E. Miller, W. Dichtel, A.K. Leibovich, A.L. Feig, J.D. Martin, K.S. Bjorkman, Z.D. Schultz, and T.L. Smith. 2015. "University Learning: Improve Undergraduate Science Education." Nature 523:282–284. https://doi.org/10.1038/523282a.

Executive Office of the President. 2022. "Space Weather Research-to-Operations and Operation-to-Research Framework." Space Weather Operations, Research, and Mitigation Subcommittee Committee on Homeland & National Security of the National Science & Technology Council, https://www.whitehouse.gov/wp-content/uploads/2022/03/03-2022-Space-Weather-R2O2R-Framework.pdf.

Moldwin, M., and C. Morrow. 2021. Solar and Space Physics Decadal Survey AIP Demographic Study Results. Ann Arbor: University of Michigan. https://deepblue.lib.umich.edu/handle/2027.42/166102.

NASA (National Aeronautics and Space Administration). 2021. Space Weather Science and Observation Gap Analysis for the National Aeronautics and Space Administration (NASA): A Report to NASA's Space Weather Science Application Program, a report to NASA's Space Weather Science Application Program, compiled by APL September 2020 to April 2021. https://science.nasa.gov/science-pink/s3fs-public/atoms/files/GapAnalysisReport_full_final.pdf.

NASEM (National Academies of Sciences, Engineering, and Medicine). 2021. Planning the Future Space Weather Operations and Research Infrastructure: Proceedings of a Workshop. Washington, DC: The National Academies Press. https://doi.org/10.17226/26128.

NASEM. 2022. Advancing Diversity, Equity, Inclusion, and Accessibility in the Leadership of Competed Space Missions. Washington, DC: The National Academies Press. https://doi.org/10.17226/26385.

NOAA (National Oceanic and Atmospheric Administration). 2021. "NOAA Announces Appointees to New Space Weather Advisory Group." News Around NOAA. September 14. https://www.weather.gov/news/091421-swag-members.

NSF (National Science Foundation). 2017. 2016 Doctorate Recipients from U.S. Universities. Alexandria, VA: National Center for Science and Engineering Statistics. https://www.nsf.gov/statistics/2018/nsf18304.

Seymour, E., and A.-B. Hunter. 2019. Talking About Leaving Revisited: Persistence, Relocation, and Loss in Undergraduate STEM Education. Cham, Switzerland: Springer Nature.

Weiss, J.-P., W.S. Schreiner, J.J. Braun, W. Xia-Serafino, and C.-Y. Huang. 2022. "COSMIC-2 Mission Summary at Three Years in Orbit." Atmosphere 13(9):1409. https://doi.org/10.3390/atmos13091409.

2

Research, Observation, and Modeling Needs:
The Sun and Heliosphere

Key Themes

In the sessions devoted to research, observation, and modeling needs related to the Sun and heliosphere, the following key themes emerged from the presentations and discussions:

- The Heliophysics System Observatory (HSO) has not been strategically planned from a systems-science perspective. As each mission in the HSO is evaluated on that individual mission's science goals, any systems science research is done on an ad hoc basis.
- In all areas of space weather, addressing the entire system of systems is important for understanding and predicting the system's behavior.
- The single most important measurement for predicting and understanding the state of the corona and its time evolution is global, high time cadence monitoring of the photospheric magnetic field.
- In order to understand and validate models of coronal mass ejections, it is important to track their evolution and propagation in three dimensions as they approach Earth.
- Continuous, global monitoring of the solar corona and near-Sun heliosphere state is important.
- Inter-calibration of the observing instruments and robust methods for incorporating data into models are crucial elements in forming the important global view of the Sun and the heliosphere from multiple vantage points.

A large share of the workshop's panels was devoted to the research, observation, and modeling needs for various aspects of space weather, from the Sun and heliosphere to the magnetosphere and ionosphere–thermosphere–mesosphere (ITM) and on to the ground effects of space weather. These presentations specifically addressed three different parts of the statement of task:

- *Examine trends in available and anticipated observations, including the use of constellations of small satellites, hosted payloads, ground-based systems, international collaborations and data buys, that are likely to drive future space weather architectures; review existing and developing technologies for both research and observations.*

- *Consider the adequacy and uses of existing relevant programs across the agencies, including NASA's Living With a Star (LWS) program and its Space Weather Science Application initiative, the National Science Foundation's (NSF's) Geospace research programs, and the NASA-NOAA-NSF research-to-operations (R2O) and operations-to-research (O2R) programs for reaching the goals described above.*
- *Consider how to incorporate data from NASA missions that are "one-off" or otherwise non-operational into operational environments, and assess the value and need for real-time data (for example, by providing "beacons" on NASA research missions) to improve forecasting models.*

This chapter and the next two summarize discussions on those topics. Here in Chapter 2, the focus is on research, observation, and modeling needs for the Sun and heliosphere. Chapter 3 focuses on the research, observation, and modeling needs for the magnetosphere, ionosphere, thermosphere, and mesosphere, while the focus of Chapter 4 is the research, observation, and modeling needs to understand the ground effects of space weather.

In summarizing the presentations of the various panels, varying approaches are used in this and the other two chapters. In some cases the summary is done in a more traditional form, describing what each panelist said in turn. In other cases, in which there was significant overlap among the panelists' presentations and the panel's output became a group effort, the session summary is more of a synthesis and has only a few comments that are attributed to individual speakers.

Several sessions during the workshop were devoted to the issue of research, observation, and needs related to the Sun and heliosphere. Four keynote speakers reviewed various research needs, mainly relating to the Sun and heliosphere. A panel of presenters addressed model and observation needs related to the Sun, and a second panel took on the topic of model and observation needs related to the solar wind.

KEYNOTES

The four keynote speakers were Drew Turner of the Applied Physics Laboratory of Johns Hopkins University, Judy Karpen of NASA's Goddard Space Flight Center, Noé Lugaz of the University of New Hampshire, and Kathryn Whitman of NASA's Johnson Space Center. The speakers were given the following key questions for consideration:

- What do we need to understand to enable predictive capability of the following topics (shown below)?
- What are the research needs to make progress on that understanding?

Turner and Karpen were asked to discuss the state of space weather science and the adequacy of and gaps in existing space weather–related programs at the National Aeronautics and Space Administration (NASA), the National Science Foundation (NSF), and the National Oceanic and Atmospheric Administration (NOAA). Lugaz and Whitman were asked to address the following questions: "What do we need to understand to predict ICMEs (including stealth) and IMF Bz?" and "What do we need to understand to predict 'all clear' SEP periods?"

Turner's talk focused on the key findings of a gap analysis conducted by the Johns Hopkins University Applied Physics Laboratory for NASA. The explicit task from NASA specified space-based observations can only be considered in the context of improved space weather predictive capabilities and the science of space weather (NASA 2021). Turner defined space weather as the intersection between the natural space environment as an object of fundamental scientific research and the threats posed to humans and technology systems that require risk mitigation by operational communities. Space weather activities span applied science and engineering as well as research-to-operations/operations-to-research (R2O2R) activities

The gap analysis Turner described breaks space weather effects into five categories: (1) direct drivers from the Sun itself, including solar flares and radio bursts; (2) radiation effects; (3) ionospheric effects, which affect radio signal propagation; (4) thermospheric expansion, which affects paths of satellites and other objects through the thermosphere; and (5) geomagnetically induced currents (GICs), which affect critical infrastructure on the ground. The NASA report characterized various space weather hazards according to their likelihood and their consequences, and it then combined that risk analysis with other factors, such as scientific merit, to prioritize current space weather observation gap categories. The top priorities in the resulting list were (1) solar and solar wind observations, including observations off the Sun–Earth line; (2) ionospheric key observables; (3) the solar wind in peri-geospace; (4) thermospheric key observables including ionospheric D- and E-region energetic particle precipitation and E- and F-region cusp and auroral precipitation; (5) ring current and radiation belt electrons; and (6) plasma sheet electrons and injections (or bursts) from cislunar into geosynchronous orbit (GEO) and medium Earth orbit (MEO) regions. Noting the importance of ground-based assets, Turner reminded the audience that the gap analysis was tasked to assess only space-based assets.

Some of the highest-priority observational gaps could be addressed with a near-future system-of-systems approach, through combining a network of state-of-the-art observatories, supporting infrastructure, and the space weather centers that develop the forecasts (see Figure 2-1 for persistent observations needed to fill the gaps). For example, to complement L1 monitors Turner suggested dedicated L4 and L5 monitors that could provide more observations of the solar disk, solar corona, and solar wind. The trailing L5 monitors are more useful for measuring potentially geoeffective active regions and solar wind stream structures,

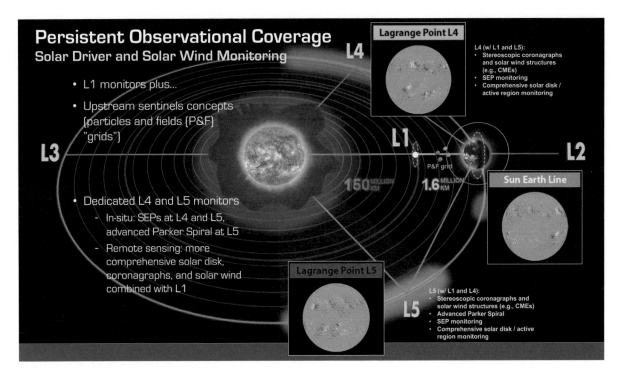

FIGURE 2-1 Persistent observational coverage of solar drivers and solar wind. Specifically, having monitors at Lagrange points 1, 4, and 5 would aid in SEP monitoring and provide more comprehensive solar disk coverage.
NOTE: CME = corona mass ejection; SEP = solar energetic particle.
SOURCE: Drew Turner, Johns Hopkins University Applied Physics Laboratory, presentation to workshop, April 11, 2022.

while the leading L4 monitors are more useful to monitor the solar energetic particles. The backside of the Sun L3 observations offer a number of distinct advantages for longer-term forecasting but are challenged by complicated communication requirements. Turner also suggested additional solar wind monitors between L1 and the magnetospheric boundary, which could provide better magnetospheric specification. Finally, persistent observations of the radiation environment and of the lunar and cis-lunar environments are needed for astronaut and infrastructure safety. In addition to the above, he said, comprehensive ionosphere, thermosphere, and radiation belt monitoring and science are needed in order to improve space weather predictions. Turner also noted that within the next decade, it is expected that the thin shell in the thermosphere will host around 10,000 spacecraft. While they will need accurate space weather information, they could also be used to provide more knowledge of this region.

Turner said that filling these observational gaps will require a systems approach with coordinated, concurrent measurements. Creating such a system will require a long-term strategy coordinated over agencies as well as dedicated and supported implementation plans. Observing system simulation experiments (OSSEs) can be invaluable in providing cost–benefit analyses with respect to model performance (e.g., the L3/L4/L5 observations of solar magnetic field versus investments into transport models and helioseismology from the far-side). Tools provided by data assimilation techniques and machine learning can be used to fill in the data gaps.

In the next talk, Karpen said that to improve space weather predictions it will be necessary to better understand the relevant physical phenomena (i.e., the solar wind structure and turbulence; the nonlinear response of the magnetosphere; the ionospheric conductances, winds, and outflows; and the coupling between the neutral atmosphere and ionosphere). Furthermore, researchers will need to figure out how to couple both regimes (e.g., solar wind to magnetosphere) and physical transitions (e.g., magnetic field to plasma-driven dynamics, sub- to super-Alfvénic, collisional to collisionless plasmas) across spatial and temporal scales. She also suggested that space weather would benefit from collaboration with plasma physics researchers, who are already taking advantage of numerical and computational advances, and with Earth scientists, whose atmospheric models are important for understanding how space weather is driven from below.

Karpen argued that idealized models will need to be advanced through the use of data assimilation and model coupling as well as through better data-driven models and those incorporating continuous coverage of the solar magnetic field. The models should use the advances made in faster and larger massively parallel computational systems. The Community Coordinated Modeling Center (CCMC) and the International Space Weather Action Team (ISWAT) are making progress on standardizing the validation and verification of research models prior to transition. However, there is a serious gap in training and employing new computational physicists in heliophysics. If space weather research is to make advances that help improve space weather forecasting, cooperation between NASA and NOAA will be crucial. Undertaking the complex task to couple Earth system models with the space weather models will also require additional funding.

Lugaz started from the notion that knowledge of the full plasma and magnetic field measurements upstream of Earth is needed in order to forecast geomagnetic storms and the radiation belt enhancements. Measuring these upstream of Earth at L1 gives about 1-hour lead time. Knowing these parameters as the plasma leaves the Sun would give the longest lead times, but that is not currently feasible. Coronal mass ejections (CMEs) are the source of the strongest interplanetary magnetic field (IMF) and thus the most critical for space weather forecasts. Other magnetic field–driven impacts are typically created by corotating interaction regions (CIRs) or more complex and very difficult to predict events, such as CME–CME or CME–CIR interactions. Ordered according to decreasing lead time and increasing accuracy, the primary tools for forecasting and prediction are modeling and simulations, remote sensing observations, and in situ measurements (see Figure 2-2).

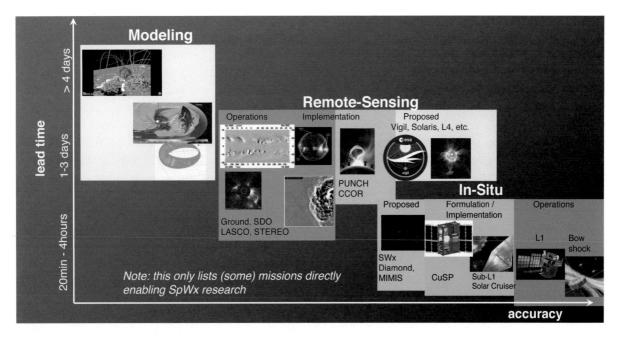

FIGURE 2-2 Lead time versus accuracy of forecasts using modeling and simulations, remote-sensing observations, and in situ measurements.
NOTE: Acronyms defined in Appendix D.
SOURCE: Noé Lugaz, University of New Hampshire, presentation to workshop, April 11, 2022.

In terms of modeling and simulations, the key questions revolve around the ways that active regions emerge and evolve and around which active regions and coronal structures result in (fast) CMEs and how and when. These questions can be addressed by different approaches, depending on whether the forecast is made before or after the eruption has occurred, but in all cases computational limitations require using a combination of full physics description and approximations.

Long (over 4 days) lead time predictions are based on physics-based simulations driven by solar observations; better accuracy with lead times of 1 to 3 days can be obtained by combining remote sensing observations either with physics-based models using solar observational inputs or with empirical models. However, complex events, superstorms, and possibly solar maximum periods in general are all likely to require full simulations that incorporate multiple aspects that are currently non-standard; for example, variable solar wind, successive CMEs and CME interactions, or other unusual conditions.

Lugaz stated that remote sensing observations are key inputs to all modeling and simulations. The key science questions include the formation and variability of CIRs, CME evolution, and more complex events such as CME–CME interactions and stealth CMEs. In order to make progress in this area, two or three distinct observational views are needed, with a view from L5 (trailing the Sun–Earth line) being particularly valuable. Observations of the polar magnetic fields (which are best done out of the ecliptic plane) are key to understanding high-speed solar wind and the related non-CME geoeffective events, as well as understanding the CME sheath regions and CME deflection.

Regarding in situ measurements, Lugaz suggested merging different approaches, including data assimilation (of both remote sensing and in situ observations), ensemble simulations, parameterization of simulations, and machine learning techniques. However, the most accurate way to forecast the IMF polarity affecting Earth is to measure it in situ in the solar wind upstream of Earth. Such observations have been primarily done at L1, but multiple measurements are needed to accurately characterize CMEs and other solar wind structures.

The key science questions that can be advanced using in situ observations involve mesoscales, for example, evolution of the CMEs and the solar wind on time scales of less than 1 day or over the 1-hour travel time from L1 to Earth's bow shock, and determining the relevant length-scales of CMEs and other solar wind structures. Prior studies have shown that at times, there can be significant differences in the solar wind measured at L1 and at Earth's bow shock. Addressing these issues requires multi-spacecraft measurements inside L1; rideshares and small satellites would be perfectly suited for these investigations. Unfortunately, as several speakers pointed out, the lack of coordination and systemic approach between missions results in only random conjunctions.

In the panel's final presentation, Whitman focused on forecasting solar energetic particle (SEP) events as well as "all-clear" conditions (those when SEP fluxes are sufficiently low to allow extravehicular or other activities impacted by space radiation). She noted that operations on the International Space Station (ISS) take place at low Earth orbit inside Earth's protective magnetosphere, which reduces the SEP threat to individuals. However, with planned missions farther out in space, prediction of both SEP events and all-clear SEP periods will become increasingly important. Furthermore, she said, the definition of "all-clear" is not consistent across the community. NOAA requirements are based on particle fluxes >10 MeV, while the Space Radiation Analysis Group (SRAG) augments that with an additional requirement for >100 MeV particles; research groups apply a variety of threshold levels to multiple energy ranges. She noted that a set of standard SEP "all-clear" definitions would ensure better usability for operations as well as facilitate cross-model validation and comparisons.

A major difficulty in predicting SEP events is that the largest events are associated with M- and X-class flares and fast CMEs, which occur only rarely; thus, the data sets are too sparse for statistical modeling. The pre-eruption forecasting is done by combining the likelihoods that an M- or X-class flare will occur, that a fast and wide CME will be produced, and that an SEP event will be produced and propagate to the location of interest. Thus, improvements in forecasting any of these phenomena should increase the reliability of SEP forecasts (Figure 2-3).

Post-eruption forecasts use information about flares, CMEs, radio waves, electrons, and protons in various combinations to estimate SEP characteristics before the particle intensities increase. Unfortunately, most models are currently unable to make predictions prior to the onset of a well-connected SEP event, and post-eruption forecasts suffer from data latency issues, manual determination of the CME parameters, and occasionally analysis run time. Machine learning holds a lot of promise in improving forecasts, particularly for pre-eruption models, Whitman said.

Whitman's final comments reflected a common theme in the session—the importance of understanding the entire system of systems and the need for high-cadence coronagraphs and magnetographs, extreme ultraviolet imaging, and in situ magnetic field, and energetic particle measurements at L1, L4, and L5 (Figure 2-4). Whitman closed by saying, "It is critical that operationally supported, high-cadence, reliable and accurate space weather data streams for all phenomena relevant to SEP production are publicly available for operations and the deployment and development of models that require real-time observations."

THE SUN

The Solar Panel, moderated by committee member Pete Riley, consisted of panelists Todd Hoeksema of Stanford University, Sarah Gibson of the University Corporation for Atmospheric Research, Cooper Downs of Predictive Science Inc., Craig DeForest of the Southwest Research Institute, Valentin Pillet of the National Solar Observatory, and Phil Chamberlin of the University of Colorado.

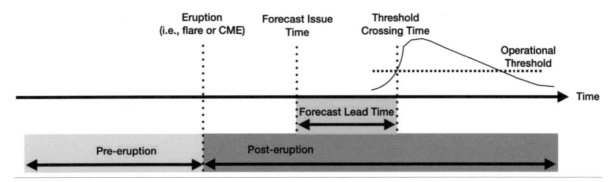

FIGURE 2-3 Illustration of pre- and post-eruption forecasting lead time.
NOTE: CME = coronal mass ejection.
SOURCE: Katie Whitman, NASA Johnson Space Center, presentation to workshop, April 11, 2022.

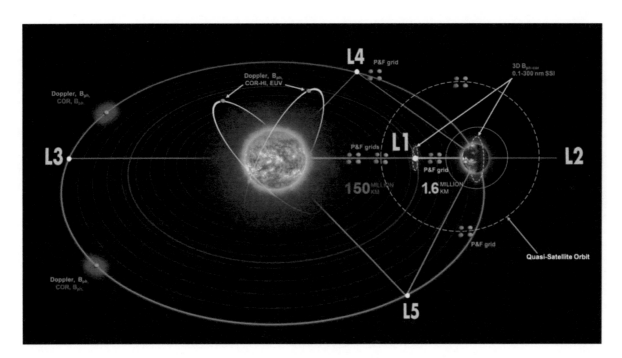

FIGURE 2-4 Locations of measurements to prioritize for space weather forecasting.
NOTE: EUV = extreme ultraviolet; P&F = Pepperl & Fuchs; SSI = Synchronous Serial Interface.
SOURCE: Katie Whitman, NASA Johnson Space Center, presentation to workshop, April 11, 2022.

A common theme from the panel was that the single most important measurement for predicting and understanding the state of the corona and its temporal evolution is global (i.e., 4π steradian), time-dependent monitoring of the photospheric magnetic field, which provides the magnetic boundary condition of the solar atmosphere. This is how all models of the solar corona and solar wind are driven, both space weather/predictive models and first-principles/magnetohydrodynamics models. The time-dependent state of the corona affects the formation of the solar wind, stream interaction regions, and CMEs. This inner boundary condition is also needed for physical understanding of fundamental processes, such as magnetic energy storage and release. Current observational views of the polar magnetic field are limited, and observations from the far side of the Sun are missing. Current full-surface boundaries are created either by a series of static snapshots over 2 weeks to view the full solar disk or else by invoking models that artificially evolve the observational data to the part of the Sun that is not visible. Thus the boundary is not well represented, the time dependence of the solar corona is not well constrained, and global models are not based on any true representation of the solar surface at any given time (Figure 2-5).

Polar magnetic field measurements of the Sun are critical to understanding the solar wind and the magnetic topology into which a CME propagates. Unlike the backside of the Sun, which eventually rotates around to the front, the polar magnetic fields have never been measured without a severely limiting observing angle. Best estimates of the polar magnetic fields are provided by the Hinode spectropolarimeter, which, however, are only sporadically available. Instrumental limitations of other full-disk data (e.g., from the Solar Dynamics Observatory's Helioseismic and Magnetic Imager [SDO/HMI] and the Global Oscillation Network Group [GONG]) do not provide consistent estimates of polar magnetic fields, and the resulting range of credible estimates is large, which then produces vastly different predicted heliospheres. Even measurements from the Solar Orbiter will not fully meet the observational needs, as measurements would be needed from both poles simultaneously.

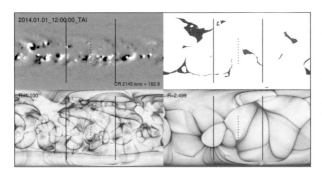 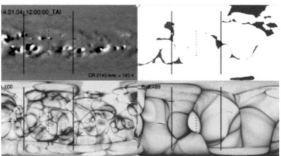

FIGURE 2-5 Limitations on solar imaging. This pair of images from http://hmi.stanford.edu/QMap/ shows the photospheric radial magnetic field (B_r) in the upper left, and corona and heliospheric open flux and Qmaps that result in the heliosphere. These two images are separated by 3 days, showing that as new active regions emerge the whole corona and heliosphere change in response. Current missions can only measure this change from the Sun–Earth line (i.e., when an existing active region rotates around the solar limb).
SOURCES: Cooper Downs, Predictive Science, Inc., presentation to workshop, April 12, 2022; generated from the Solar Dynamics Observatory Joint Science Operations Center, "Q-Map Products Based on Potential Field Source Surface Coronal Field Model," http://hmi.stanford.edu/QMap; courtesy of the Stanford Solar Observatories Group and Predictive Science, Inc.

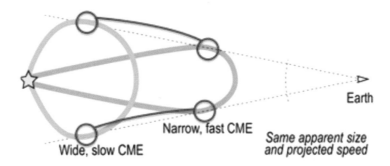

FIGURE 2-6 The ambiguity between whether a coronal mass ejection (CME) is wide and slow, or narrow and fast as seen from a single Earth viewpoint.
SOURCE: Craig DeForest, Southwest Research Institute, presentation to workshop, April 12, 2022.

Another common theme of the panel was the need to track the evolution of CMEs in three dimensions as they travel toward Earth. As solar wind and CME structures evolve between the Sun and Earth, current L1 coronagraph measurements are incapable of distinguishing between narrow, fast CMEs and wide, slow CMEs (Figure 2-6). Panelists suggested several possible observing scenarios, including multiple viewpoints (e.g., stereoscopy between an L1 measurement and off-Sun–Earth-line measurements) of a CME during its propagation. Several speakers mentioned the value of a polar coronagraph and heliospheric imagers, as the polar view provides a top view for all CMEs traveling near the ecliptic plane where the planets reside and allows detection of the longitudinal structure of CMEs, corotating interaction regions, and shocks. A near-term solution will be offered by the Polarimeter to Unify the Corona and Heliosphere (PUNCH) mission, which will use polarized white-light coronagraphs and heliospheric imagers to provide three-dimensional information and the chirality of the CME magnetic field, which will provide information of the IMF polarity upon arrival at Earth. The ratio of polarized light to total brightness provides information on the distance of the plasma feature from the observer; time series of this ratio will then provide three-dimensional information on the plasma structure.

The panel pointed out that most current CME propagation models do not include the magnetic field. Magnetohydrodynamic (MHD) models include the magnetic field but are computationally expensive given the large system size and resolution needed to resolve the structure in scales relevant to Earth. Modeling magnetized CMEs requires starting with an appropriate estimation of the magnetic field at the solar boundary, which has its limitations discussed above. The ideal solution may be a combined modeling and data assimilation approach that has been successfully used in Earth sciences applications, such as hurricane tracking.

In addition to the global photospheric magnetic field, global imaging of the middle corona was identified as an important missing piece to understand the Sun. For example, understanding the global restructuring of the corona after a CME requires global imaging of the entire corona (i.e., measurements from different viewpoints). In addition to the vantage point issues, the middle corona (1.5-4 solar radii) represents an observational gap in the range of imagers from extreme ultraviolet (EUV) monitoring the solar surface and white light coronagraphs recording the outflowing plasma at higher altitudes. As this gap coincides with the critical region where CMEs are accelerated, it creates uncertainties in physical models of CME initiation. Future solar imagers should eliminate such sampling gaps either by extending the fields of view of EUV imagers higher or by reducing the noise in white light measurements close to the solar surface.

A longer-term goal is to develop the capability to routinely produce maps of the vector magnetic field in the photosphere, chromosphere, and corona. This would be a significant advancement over the current situation where the fields above the photosphere are inferred from models. In particular, observations of filament channels and magnetic neutral lines would allow observing the energy storage and release processes. Furthermore, measuring the chirality in the chromosphere at magnetic neutral lines would help track magnetic helicity build-up and provide knowledge of the characteristics of solar eruptions upon liftoff.

Understanding the global solar dynamo processes will not only allow better understanding of the overall 11-year solar cycle, it will also allow assessment of where and when "active nests" on the solar surface will occur, how much flux will appear in an emerging region, how the emerging flux will interact with existing flux patterns, or how complex the structure will be. Progress on these topics requires monitoring the solar interior, which can only be done through indirect helioseismology measurements; this would also benefit from multiple viewpoints.

The solar cycle and solar dynamo are important for the long-term space weather predictability (how active will the next solar cycle be?) and the coupling of space weather to atmospheric weather and climate. Measurements of the meridional circulation (measured via magnetic fields) can offer insights into the solar cycle because meridional circulation may act like the solar cycle "clock." However, a challenge for the observations is that these phenomena should be measured consistently over long (solar cycle) time scales.

Solar irradiance is a key input for ITM driving and thus needs to be measured continuously. Solar EUV irradiance and soft X-rays are the primary energy input into the upper atmospheres of Earth and other planets, including Mars. As Chamberlin said, "It's the photons! Not just the plasma!" There are no specific warnings either for short-term irradiance variations caused by solar eruptive events (flares) or for longer-term solar activity variations. Yet, irradiance variations have an immediate impact on the atmosphere, especially at low latitudes. Solar eruptive events are not understood to a great enough degree to predict their impact on irradiance, proxy models do not capture spectral variations, and physics-based irradiance models are computationally expensive. Irradiance measurements from L5, complementing those from the Sun–Earth line, would provide information on evolution of active regions and help estimate the CME mass via coronal dimming measurements.

THE SOLAR WIND

The Solar Wind Panel, moderated by committee member Nicholeen Viall, followed the Solar Panel. Its panelists were Joe Borovsky of the Space Science Institute; Vic Pizzo of NOAA's Space Weather Prediction Center; Stuart Bale of the University of California, Berkeley; Nick Arge of NASA's Goddard Space Flight Center; and Erika Palmerio of Predictive Science, Inc.

Continuing the common theme, the Solar Wind Panel noted the need for continuous, global monitoring of the Sun and the solar corona. Basic understanding of the solar wind and improving the solar wind models will require solar observations, as discussed above, as well as remote sensing and in situ measurements of the solar wind itself. Global coverage will require a fleet of monitors similar to those needed for solar observations (e.g., L4, L5, far-side, and over the solar poles). The panel members generally believed that global coverage is more important than increasing spatial resolution (Table 2-1), and it was noted

TABLE 2-1 Critical Data Products Needed to Improve, Validate, and Constrain Coronal and Solar Wind Models

Critical Data Product	Impact	Current and Needed Observational Instrumentation
Global synchronic magnetic field maps	Improved B.C. used to drive coronal and space weather models, especially time-dependently	SDO/HMI and SolO/PHI Polar SolO *Missing: Polar imager, continuous far-side imaging*
Global synchronic EUV maps	Coronals holes identified in EUV maps can be used to V&C coronal models	ST A&B+SDO ST A+SDO+SolO *Missing: Polar imager, continuous far-side imaging*
Coronal 3D electron density (N_e) and plane-of-the-sky magnetic field reconstructions	3D WL electron density (N_e) tomographic reconstructions and plane-of-the-sky WL images segmented to surmise the coronal magnetic field observationally — Used to V&C models — Multiple viewpoints improve V&C	ST A and/or B+SOHO ST A+SOHO+SolO+CODEX+PUNCH *Missing: Out of the plane and widely spaced, strategically located imaging*
Multi-vantage-point in situ plasma observations	SW plasma observations from multiple, widely spaced vantage points used to V&C SW models	(L1: ACE, WIND, DSCOVR), Ulysses, STEREO A&B +PSP, SolO *Missing: Out of the plane and widely spaced, strategically located imaging*

NOTE: 3D = three-dimensional; A = ahead; ACE = Advanced Composition Explorer; B = behind spacecraft; B.C. = boundary conditions; CODEX = Coronal Diagnostic Experiment; DSCVR = Deep Space Climate Observatory; EUV = extreme ultraviolet; Ne = electron density; PHI = Polarimetric and Helioseismic Imager; PSP = Parker Solar Probe; PUNCH = Polarimeter to Unify the Corona and Heliosphere; SDO/HMI = Solar Dynamics Observatory's Helioseismic and Magnetic Imager; SOHO = Solar and Heliospheric Observatory; SolO = Solar Orbiter; ST = STEREO = Solar Terrestrial Relations Observatory; SW = Solar wind; V&C = validate & constrain; WL = white light.
SOURCES: Nick Arge, NASA, presentation to the workshop; data from C.N. Arge, S. Jones, C.J. Henney, S. Schonfeld, A. Vourlidas, K. Muglach, J.G. Luhmann, and S. Wallace, 2021, "Multi-Vantage-Point Solar and Heliospheric Observations to Advance Physical Understanding of the Corona and Solar Wind," Heliophysics 2050 white papers, https://www.hou.usra.edu/meetings/helio2050/pdf/4056.pdf, and A. Posner, C.N. Arge, J. Staub, O.C. StCyr, D. Folta, S.K. Solanki, R.D.T. Strauss, F. Effenberger, A. Gandorfer, B. Heber, C.J. Henney, J. Hirzberger, S. Jones-Mecholsky, P. Kuehl, and O. Malandraki, 2021, "A Multi-Purpose Heliophysics L4 Mission," *Space Weather* 19:e2021SW002777, https://doi.org/10.1029/2021SW002777.

(see above discussion on solar observations and models) that the accuracy of the initial and boundary conditions at the Sun are critical for the accuracy of the solar wind models.

The optimal number of measurement points was discussed, with the panelists generally believing that there is a limited return on adding more than a couple of spacecraft in the ecliptic plane. While plans for wider longitudinal spread of solar and solar wind observations exist, the panel pointed out the need for high-latitude observations for both the photospheric magnetic fields and off-ecliptic views of CMEs. Thus, some panel members advocated prioritizing a polar viewpoint as opposed to additional in-ecliptic monitors. Major advances in observing polar fields and CME tracking could come from polar view missions, such as Ulysses; four spacecraft could provide continuous coverage of both poles, which would also allow tracking CME deflections, rotations, deformations, and interactions throughout the inner heliosphere. Furthermore, polar imagers would illuminate the extent and causes of the deviations of the IMF direction from the nominal Parker spiral, thereby improving understanding and predictions at Earth orbit. However, polar operational missions are expensive; Pizzo suggested that the missions will need to be designed to have very focused goals and expectations to manage the cost. Panelists also mentioned additional methods to track the solar wind and CMEs, specifically in situ measurements between the Sun and L1 and Faraday rotation measurements from the ground.

Two corollaries to the importance of multiple viewpoints and global coverage are that the inter-calibration of instruments on board those spacecraft is crucial and that robust methods are needed for incorporating the data into the models (see also Chapter 5). Magnetograph data, for example, vary in spatial resolution and polarization sensitivity, in the spectral lines observed, and in sampling patterns, making inter-calibrating magnetograph data a notoriously difficult task.

Global modeling of the coupled system is crucial, as understanding CMEs and SEPs requires a holistic Sun-to-heliosphere model. As CMEs and SEPs propagate within—and are transported through—the solar wind, it is necessary to have a model that simultaneously captures the states of the solar corona, the solar wind, and the CMEs. CME–CME interactions are known to influence their evolution, but the models are currently not well constrained. Furthermore, the lack of quantitative measurements of the magnetic field within CMEs makes it difficult to develop a full MHD model of CME propagation for operations (without data to serve as model input, constraints, and validation points). The panel noted that the modeling and predicting of the CME magnetic field are progressing and need to start to transition to operations (including validation). Likewise, some SEP models have real-time capabilities, but they need to be tested and their performance evaluated for real-time forecasting.

Finally, accurate measurements of the solar wind that actually hits Earth's magnetosphere are needed to understand—and ultimately predict—the driving of magnetospheric dynamics. Measurements of the solar wind from L1 can deviate from those measured at the magnetopause due to triple aberration and inherent structures in the solar wind. Triple aberration (Figure 2-7) is caused by the combination of the motion

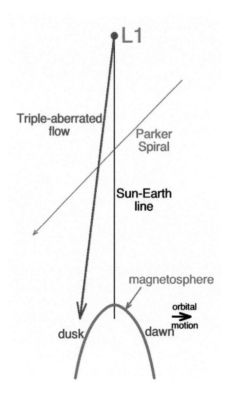

FIGURE 2-7 The triple-aberrated solar wind flow. The result of this flow is that the solar wind measured at L1 is not always the solar wind that affects the magnetosphere.
SOURCES: Joe Borovsky, Space Science Institute, presentation to workshop, April 12, 2022; from J.E. Borovsky, 2022, "The Triple Dusk-Dawn Aberration of the Solar Wind at Earth," Frontiers in Astronomy and Space Sciences, June 6, https://doi.org/10.3389/fspas.2022.917163, Copyright © 2022 Borovsky.

of Earth around the Sun, the non-radial component of the solar wind flow, and the interplay of magnetic structures, which propagate outward along the Parker spiral faster than the bulk solar wind plasma. As the solar wind flow velocity vector varies by about ±5 degrees over the 220 R_E distance from L1 to Earth, it causes a deflection that is about ± 30 R_E, or roughly the width of the magnetosphere. This would indicate that a plasma parcel going through L1 might completely miss hitting Earth's magnetosphere due to triple aberration only. The challenges can be solved by a solar wind monitor between L1 and Earth. To address the inherent structures in the solar wind, the best solution would be to develop a better physical understanding of their causes. For example, it is not currently known how much of the structure is created at the Sun as the solar wind is formed, versus how much is generated en route through turbulence. To best disentangle the origin of the solar wind structures, improved measurements of elemental composition should be leveraged to differentiate between the structures created at the Sun versus structures created during propagation through the heliosphere (see also the section on new architectures in Chapter 6).

REFERENCES

Arge, C.N., S. Jones, C.J. Henney, S. Schonfeld, A. Vourlidas, K. Muglach, J.G. Luhmann, and S. Wallace. 2021. "Multi-Vantage-Point Solar and Heliospheric Observations to Advance Physical Understanding of the Corona and Solar Wind." *Heliophysics 2050 White Papers*, 4056. https://www.hou.usra.edu/meetings/helio2050/pdf/4056.pdf.

NASA (National Aeronautics and Space Administration). 2021. *Space Weather Science and Observation Gap Analysis for the National Aeronautics and Space Administration (NASA): A Report to NASA's Space Weather Science Application Program*. Washington, DC. https://science.nasa.gov/science-pink/s3fs-public/atoms/files/GapAnalysisReport_full_final.pdf.

Posner, A., C.N. Arge, J. Staub, O.C. StCyr, D. Folta, S.K. Solanki, R.D.T. Strauss, et al. 2021. "A Multi-Purpose Heliophysics L4 Mission." *Space Weather* 19:e2021SW002777. https://doi.org/10.1029/2021SW002777.

3

Research, Observation, and Modeling Needs: Magnetosphere, Ionosphere, Thermosphere, and Mesosphere

---

Key Themes

In the sessions devoted to the research, observation, and modeling needs of the magnetosphere and the ionosphere–thermosphere–mesosphere (ITM), the following key themes emerged from the presentations and discussions:

- In all areas of space weather, dealing with the entire system of systems is important to understanding and predicting its behavior.
- There is a lack of understanding of extreme space weather conditions, their drivers and impacts, and further measurements are needed on these rare events.
- Multipoint measurements in an arrangement providing global coverage are key in resolving the space weather processes. Inter-calibration of the observing instruments as well as development of robust methods for incorporating the data into models are crucial elements of success.
- Earth's magnetosphere needs to be studied as a system of systems, with focus on both better observational coverage (satellites and ground-based) and modeling of the mesoscale processes.
- Understanding the ionosphere and thermosphere system and the couplings from both above and below is needed to predict space weather. Although multi-scale and multi-step physical processes are important in understanding the ITM system and space weather, they are neither well observed nor well modeled. In particular, forcing of the near-Earth plasma by the atmosphere from below is an underappreciated aspect of space weather.

Four of the workshop's panels were focused on research, observation, and modeling needs for the magnetosphere and the ionosphere–thermosphere–mesosphere (ITM). These panels were the Observational and Modeling Needs for the Magnetosphere Panel, the Research Needs for the Ionosphere-Thermosphere Panel, the Observation and Modeling Needs for the Ionosphere-Thermosphere Panel, and the Research Needs for Cross-Scale and Cross-Regional Coupling Panel. As previously noted, these presentations specifically addressed three different parts of the statement of task:

- *Examine trends in available and anticipated observations, including the use of constellations of small satellites, hosted payloads, ground-based systems, international collaborations and data buys, that are likely to drive future space weather architectures; review existing and developing technologies for both research and observations.*

- *Consider the adequacy and uses of existing relevant programs across the agencies, including NASA's Living With a Star (LWS) program and its Space Weather Science Application initiative, the National Science Foundation's (NSF's) Geospace research programs, and NOAA's Research to Operations (R2O) and Operations to Research (O2R) programs for reaching the goals described above.*
- *Consider how to incorporate data from NASA missions that are "one-off" or otherwise non-operational into operational environments, and assess the value and need for real-time data (for example, by providing "beacons" on NASA research missions) to improve forecasting models.*

The presentations are summarized in the following sections. Those summaries take different approaches depending on the panel. In some parts, the summary described what each panelist said in turn. In other cases, however, for panels in which there was significant overlap among the panelists' key messages and the panel's output became a group effort, the session summary is more of a synthesis and has fewer comments attributed to individual speakers.

THE MAGNETOSPHERE

The Magnetosphere Panel was moderated by committee member Terry Onsager. Its panelists were Christine Gabrielse of the Aerospace Corporation, Larry Kepko of NASA's Goddard Space Flight Center, Matina Gkioulidou of the Applied Physics Laboratory of Johns Hopkins University, and Vania Jordanova of Los Alamos National Laboratory. The panelists were asked to address two key questions:

- What do we need to understand to enable predictive capability of the magnetospheric state and irregularities?
- What are the research needs required to make progress on that understanding?

Several of the panel members expressed that the magnetosphere needs to be measured and modeled as a system of systems with an emphasis on the mesoscale structures and processes that control its dynamics. Kepko noted that mesoscales are the messengers and connectors of energy and mass across regions (e.g., tail to inner magnetosphere) and across scales (kinetic to mesoscale to global). Mesoscales in the magnetosphere can be defined as roughly from 1,000 km to a few Earth radii, and about hundreds of meters to 1,000 km in the ionosphere. Measuring and modeling this cross-regional and cross-scale coupling is an important missing element for space weather (highlighted also in Session 2 discussion of cross-scale and cross-regional processes). It was pointed out that such an observational program requires a system-of-systems approach (see also Drew Turner talk in New Research Needs). In addition, data assimilation tools need to be combined with models that bridge the micro- and macroscale processes to address key problems, such as the rapid particle injection and acceleration during substorms, plasma wave generation and feedback on the particle populations, and magnetosphere–ionosphere coupling though particle precipitation.

Kepko emphasized, supported by several other panelists, that comprehensive measurements of Earth's magnetosphere as a system of systems will require observational capabilities that are coordinated by design, not as an afterthought (Figure 3-1). Current data sets are sparse and do not adequately capture the mesoscales. Measurements have been carried out at small/kinetic scales and at global scales, but not at mesoscales. Resolving mesoscale processes will require coordinated satellite constellations combined with imaging components. Such an ambitious space program will need coordination with ground-based observations as well as with international, interagency, and industrial partners (see also the New Architectures session in Chapter 6).

The Heliophysics System Observatory (HSO) is a valuable resource that provides coordinated observations in pursuit of system science. However, the current coordination is ad hoc and often accidental, and it is inadequate for resolving the mesoscale. Earth's magnetosphere has an aging, inadequate fleet of satellites, and nothing is currently in the queue either for science or for space weather monitoring. The inadequacy of the available measurements is clear when considering the structures and dynamics indicated by global numerical models (Figure 3-1). The properties of the dynamic structures seen in the models cannot be measured with currently available three or four widely spaced satellites, and it is therefore not possible to adequately constrain or validate the models with existing data.

The various systems that make up the magnetosphere (bow shock, magnetosheath, magnetopause, plasmasphere, ring current, radiation belts, magnetotail, etc.) cannot be studied in isolation. The properties of these systems are "emergent"; that is, they emerge through interactions with the other systems. These interactions typically occur at the mesoscales. For example, when reconnection occurs on the dayside magnetopause, that information is communicated throughout the magnetosphere by flux transfer events, which are mesoscale structures. And when the magnetotail subsequently goes unstable and releases a substorm, that configuration change is transferred to the inner magnetosphere and ionosphere through mesoscale structures.

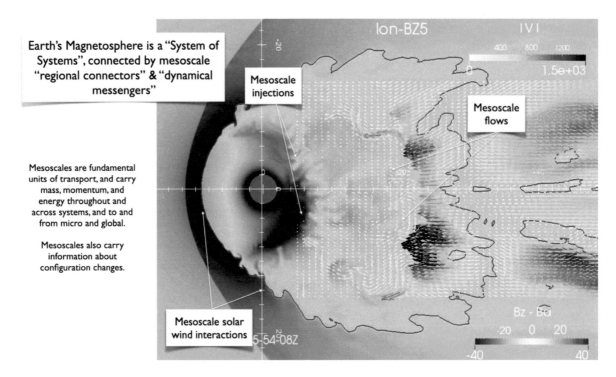

FIGURE 3-1 A model representation of mesoscale processes in Earth's magnetosphere. Observing such localized processes and understanding their global impact requires monitoring the entire system at mesoscales.
SOURCES: Larry Kepko, NASA Goddard Space Flight Center, presentation to workshop, April 12, 2022; adapted from Supporting Information Movie S1 at 1995-03-21T05-54-08Z in M. Wiltberger, V. Merkin, J.G. Lyon, and S. Ohtani, 2015, "High-Resolution Global Magnetohydrodynamic Simulation of Bursty Bulk Flows," *Journal of Geophysical Research* 120:4555–4566, https://doi.org/10.1002/2015JA021080; © 2015 American Geophysical Union, all rights reserved.

To achieve the required observing infrastructure, various panel members suggested taking a new coordinated international approach to provide multiple constellations of spacecraft in key regions to resolve the mesoscale structures. This would include mesoscale dynamics (e.g., flow burst and magnetopause flux transfer events), inner magnetosphere/magnetotail coupling, the cold plasma distribution, auroral coupling, ion outflow, and the coupling of kinetic processes to mesoscale structures. Geospace state variables, such as the upstream solar wind, auroral configuration, cross-polar-cap potential, the state of the radiation belts, polar cap open flux, solar irradiance, cold plasma density, auroral field–aligned currents, ionospheric convection maps, geomagnetic indices, and ionospheric total electron content need to be measured. The space-based measurements must be coordinated with ground-based measurements, and the full set of observations must in turn feed into and validate advanced numerical models and data assimilation. All of these activities should focus on studying the geospace holistically as a system and at the scale sizes we now know are driving the system dynamics.

One key region where mesoscale processes are particularly important is the transition region in the magnetotail between the stretched magnetic field in the plasma sheet and the dipolar magnetic field closer to Earth. As plasma flows rapidly earthward from the magnetotail reconnection site, it encounters the strong, rigid magnetic field near Earth, which causes considerable dynamics, energy conversion, and transport that occur over mesoscale (roughly 1 to 4 Earth radii) distances.

Previous satellites have sampled one or a few simultaneous locations within this region during dynamical time periods, but this has not been sufficient to resolve the mesoscale structures. Some members of the panel suggested it would be useful to have a grid of satellites with magnetometers, particle detectors, and potentially other instruments combined with ground-based and space-based imagers that can measure the two-dimensional structure and dynamics. The measurements available from such a collection would be essential in developing, driving, and validating current (research and operational) models.

Detailed knowledge is also needed of the near-Earth plasma ion composition. For example, the presence of N+ ions plays a key role in ionospheric outflow (Ilie 2021). However, information about the ion composition mostly comes from decades-old measurements. This observation gap inhibits the ability to accurately predict and characterize hazardous space weather events in the near-Earth space environment. As detailed in recent Heliophysics 2050 white papers, significant technological advances now enable more sophisticated mass spectrometers in smaller design packages than previously possible (Fernandes et al. 2021).

Another gap in current observations is the cold plasma distribution and variability. Both fundamental research and space weather predictions require the inclusion of the plasmaspheric dynamics, as the cold plasma affects many magnetospheric processes, such as the properties of the plasma waves controlling the behavior of the radiation belt electrons and the ring current ions. The cold particle populations with energies less than 100 eV are difficult to measure and therefore are the least studied, but they often dominate the plasma density. Recent Heliophysics 2050 white papers detailed the need to understand cross-scale dynamics of the plasmasphere (Goldstein et al. 2021) and the innovations needed to measure the cold plasma populations in space (Delzanno et al. 2021).

SmallSats were described as a promising approach to allow a larger number of satellites at an affordable cost. GTOSat (Blum et al. 2020), a NASA Goddard 6U CubeSat led by principal investigator Dr. Lauren Blum, is one such pathfinder. Originally planned for launch in summer 2022[1] to a low inclination geosynchronous transfer orbit, it will carry a magnetometer and a relativistic electron and proton detector that weighs only 1 kg. As this and other missions demonstrate lower-cost capabilities, the opportunities to obtain large numbers of simultaneous measurements within the magnetosphere will increase in the near future.

---

[1] At the time of publication, GTOSat had not been launched.

Two-dimensional images of the magnetosphere were described as an important complement to in situ space plasma measurements. Mesoscales, which are roughly one to a few Earth radii in size, are difficult to constrain with even three or four satellites. Images can provide the essential spatial context for multi-point, in situ spacecraft observations. Constellations of satellite measurements in the magnetosphere and ionosphere need to be coordinated with high-resolution two-dimensional images that resolve the meso-scales. For example, energetic neutral atom imaging, auroral imaging, and plasmasphere imaging can all monitor the dynamics of the geospace system in the key transition regions.

Recent studies using energetic neutral atom imagers on the TWINS satellites and all-sky imagers from the THEMIS network have documented flow channels coming Earthward from the magnetotail before a sudden storm commencement, indicating the importance of mesoscale injections to inner magnetosphere dynamics (Adewuyi et al. 2021). Although the TWINS mission was recently decommissioned and the THEMIS all-sky imagers may only operate for a few more years, Canada is implementing an all-sky camera array. In this situation there are important questions for heliophysics, including what the U.S. infrastructure contributions will be to two-dimensional imaging of the magnetosphere and how U.S. researchers can participate in internationally coordinated efforts. Ground-based radars were also noted as an important asset for imaging mesoscale dynamics in the magnetosphere–ionosphere system.

Numerical modeling was also described as an essential component of resolving the mesoscale structures and connecting the microscale, mesoscale, and global processes. Improving space weather predictions will require models that can connect the various regional models and incorporate the mesoscale processes and the cross-scale dynamics into the global model. For example, existing magnetohydrodynamic (MHD) models have demonstrated the importance of the mesoscale process in particle transport within the magnetosphere. However, the differentiation between adiabatic and non-adiabatic energization processes and their larger-scale consequences can only be done by coupling the kinetic scale physics with other models within the magnetospheric domain.

As geospace is a system of systems, models need to connect those systems in the critical mesoscale regimes. Efforts to develop such models need to be coordinated with the observations described above and will likely require new computational schemes. In addition to bridging the macro and micro scales, the modeling effort needs to be combined with data assimilation (see Chapter 5). Although scientists strive to use the highest resolution possible in numerical models, in the case of the complex global magnetosphere, mesoscale resolution may be the most practical near-term goal. Such models must include the full range of processes from the solar wind–magnetosphere interface to the ionosphere and various layers of the neutral atmosphere. The Center for Geospace Storms is one effort to model this multiscale atmosphere-geospace environment (see Figure 3-2).

As was brought up both in Session 2 and in the Session 3 Solar Wind panel, accurate measurements of the solar wind that hits the magnetosphere are essential for developing and running accurate space weather models. Solar wind measurements are used both as an outer boundary condition for magnetosphere models as well as the input to proxy models for internal boundary conditions, such as geostationary orbit state parameters for which observations are not always available. With appropriate space weather measurements, the models can provide a forecast with up to an hour lead time. Accurate forecasts of the magnetosphere–ionosphere system require equally accurate measurements of the solar wind that actually impacts the magnetosphere. It was pointed out that L1 measurements may not be sufficient due to a variety of processes creating variability between L1 and Earth's bow shock (see also the discussions of the Solar Wind panel).

Artificial intelligence (AI) and neural networks were also highlighted as having important applications in magnetospheric modeling. A recent model by Claudepierre and O'Brien (2020) was shown to reproduce well the radiation belt electron fluxes throughout the inner magnetosphere using input data from low Earth orbit electron flux measurements and the K_p geomagnetic index.

What are the key strategic model capabilities needed to advance (a) research and (b) operations?

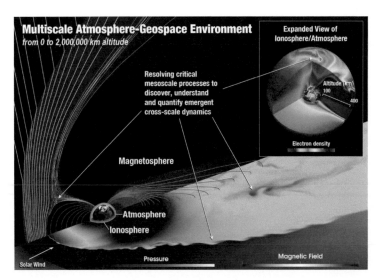

Geospace: System of Systems

We need models that connect those systems in the critical mesoscale resolution

FIGURE 3-2 Geospace modeling requires a system of systems approach due to the coupling between the magnetosphere, ionosphere, and atmosphere. All domains are linked, meaning that a holistic approach is needed to understand phenomena at all length scales.
SOURCE: Matina Gkioulidou, Johns Hopkins University Applied Physics Laboratory, presentation to workshop, April 12, 2022.

The panel also highlighted a number of valuable concepts that have been documented in the recent Heliophysics 2050 white papers[2] and that complement the issues raised by the panel. More powerful computing technologies will lead to improved physics-based modeling capabilities in areas such as multi-fluid, hybrid, and Vlasov codes as well as implicit and kinetic solvers. These advances should provide opportunities to significantly improve space weather prediction and our understanding of the geospace system of systems. The increasing amounts of data will lead to novel combinations of physics-based models with machine learning, deep learning, and "physics emulator" approaches. AI has already been successfully applied for automatic event identification, feature detection and tracking, and uncertainty quantification. These advances would benefit from collaborations with other disciplines, and therefore space weather data sets should be promoted in computer science, engineering, and other related fields.

The panel also described specific examples where advances in magnetospheric observations and modeling would directly benefit the space weather user community. One example was the need to better constrain the radiation environment within the magnetosphere to enable more efficient engineering options for satellite design. The radiation environment is an important source of solar cell voltage degradation, and current statistical models (AE9/AP9) are insufficient to predict fluences on time scales of days, months, and even a couple of years.

---

[2] See NASA Science Mission Directorate Heliophysics Division, "Heliophysics 2050 Workshop Program," Lunar and Planetary Institute and Universities Space Research Association, http://www.hou.usra.edu/meetings/helio2050/program.

User Motivation: Energy Transport Important for Ionosphere Heating

Flows, precipitation
→ Heating, upwelling
→ Effects on mesosphere/thermosphere

Upwelling → Satellite Drag

Existing models use solar wind parameters to drive the system
→ Capture large-scale global effects
→ Meso-scale phenomena are **poorly** characterized
→ Meso-scale phenomena are very dynamic!

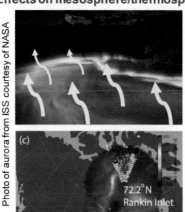

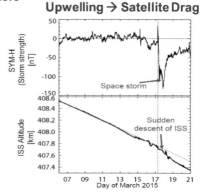

Nishimura et al., AGU talk 2016

Bottom: Gabrielse et al. (AGU JGR 2018)

FIGURE 3-3 Effects of energy transport between the magnetosphere and the ionosphere. Due to the coupled system, upwelling of the atmosphere can lead to sudden altitude changes of spacecraft.
SOURCES: Christine Gabrielese, The Aerospace Corporation, presentation to workshop, April 12, 2022. Middle top image modified from NASA Johnson Space Center, 2011, "Aurora Australis 1," video, September 17, 2011, https://eol.jsc.nasa.gov/beyondthephotography/crewearthobservationsvideos/Aurora.htm. Middle bottom image modified from C. Gabrielse, Y. Nishimura, L. Lyons, B. Gallardo-Lacourt, Y. Deng, and E. Donovan, 2018, "Statistical Properties of Mesoscale Plasma Flows in the Nightside High-Latitude Ionosphere," *Journal of Geophysical Research: Space Physics* 123:6798–6820, https://doi.org/10.1029/2018JA025440; © 2018 American Geophysical Union, all rights reserved. Right image from Y. Nishimura, presentation at the 2016 Fall Meeting of the American Geophysical Union, data obtained from the Space Physics Data Facility.

It was shown that a model using data from the two van Allen probes was able to more accurately predict the observed voltage degradation on satellites than the statistical model.[3] With additional satellites in the geostationary transfer orbit, the magnetic local time dependence of the radiation environment could also be modeled.

Another important user need for space weather information involves the atmospheric heating by energy input from the magnetosphere to the ionosphere and atmosphere (see Figure 3-3). The subsequent upwelling of the atmosphere increases satellite drag and impacts spacecraft lifetime as well as efforts to manage space traffic and collision avoidance. Existing models that are driven by solar wind parameters can capture much of the large-scale global effects, but the mesoscale phenomena are poorly characterized by current models and can be highly dynamic.

Finally, the panelists identified two concerns: The first was the need to further develop the workforce needed to advance space weather capabilities. Although space weather is a discipline with high societal importance, there seems to be a lack of career opportunities that can attract people with the right expertise

---

[3] Gabrielse et al., 2022, "Radiation Belt Daily Average Electron Flux Model (RB-Daily-E) from the Seven-Year Van Allen Probes Mission and Its Application to Interpret GPS On-orbit Solar Array Degradation," submitted to the workshop.

to enter and to stay in the field (also see the Diversity in Workforce session in Chapter 1). The second concern was the current lack of information sharing between space weather users and the scientific community. Many users are unwilling to share proprietary data, which can make it difficult for researchers to understand what information is needed to address the space weather impacts. Furthermore, it is expected that as the technological applications advance, the need for space weather information will increase, but many companies may not yet realize how they are or will be impacted by space weather. Improving communication between researchers and users of space weather information could accelerate the progress in understanding space weather and its utility.

IONOSPHERE AND THERMOSPHERE

Research Needs

The Ionosphere and Thermosphere Research Needs Panel addressed the state of the ionosphere–thermosphere and its irregularities. The five panelists were Seebany Datta-Barua of the Illinois Institute of Technology, Charles Carrano of Boston College, Jonathan Snively of Embry-Riddle Aeronautical University, Sean Bruinsma of the Space Geodesy Office of the French space agency CNES, and Greg Ginet of the Massachusetts Institute of Technology. The panelists were asked to address two key questions:

- What do we need to understand to enable predictive capability of the thermospheric and ionospheric state and irregularities?
- What are the research needs required to make progress on that understanding?

The panel's answer to the first question was that predictive capability with quantified uncertainties requires understanding of how the state of the ionosphere evolves (due to driving from above and from below) to create the irregularities, how the irregularities affect signal propagation, and what are the impacts of the changed signal propagation on the system performance. To make progress on that understanding (second question above), the panel identified global instrument network data sets that are routinely calibrated and readily available; coordination strategies across communities that share data (such as the Global Navigation Satellite System [GNSS] for Earth science and heliophysics); models with flexible interfaces and products that can be used with other models; and modeling strategies and community-adopted assessments validation metrics.

Datta-Barua began the session with a general description of the complexities of the ionospheric state, with dependencies on latitude (high-latitude, mid-latitude, and low-latitude bands) as well as scale (small ~100 m to to mesoscale ~1,000 km). The inputs to this system come from the Sun and the magnetosphere as well as from neutral dynamics, acoustic–gravity waves, traveling ionospheric disturbances (TIDs), and turbulence, all of them interacting with the geomagnetic field.

Turning to the operational applications, Datta-Barua focused on safety-critical Global Positioning System (GPS) services, such as aircraft landing systems. Aviation users require both positioning accuracy and the assurance that uncertainties due to ionospheric ranging errors are within acceptable limits, or a prompt alert to switch to other navigation methods to maintain their safety.

Better predictive capabilities would require estimates with quantified uncertainties of safety-critical parameters, Datta-Barua said. Furthermore, new methods—either deterministic or stochastic—are needed for forecasting the onset of instabilities. The forecast updates are needed on time scales of minutes to hours (as opposed to the currently available daily updates) and with spatial scales extending from global to tens of meters.

Making progress on understanding the underlying physics, Datta-Barua said, will require continuous support for geodetic GNSS (Global Navigation Satellite System) networks and upgrading these with multi-frequency sensors. Another means to sample the irregularity structures would be via space-borne measurements. However, the data records are only useful for the statistical approaches if they cover long time periods and are available and accessible to the community.

Carrano spoke about ionospheric radio wave scintillation (i.e., rapid changes in the wave propagation caused by ionospheric electron density variations that can affect the performance of various technological systems, including satellite communication and GNSS). Scintillations can be measured using ground-to-satellite and satellite-to-satellite transmissions, but gaining real-time specification and forecasts will require a fusion of ground- and space-based data sources, including the use of existing sensors and receiver networks as well as an improved theoretical understanding of the ionospheric state and the driving mechanisms that cause irregularities to develop. This in turn will require support for sensors and networks, modeling efforts, and system impact models.

Ginet addressed the effects of ionospheric disturbances on the propagation of high-frequency (HF) radio waves (i.e., 3-30 MHz). Because HF waves can reflect from the bottom of the ionosphere, they can be used for long-distance communications (e.g., AM radios) and over-the-horizon radar (OTHR) applications. However, improving the performance of these systems, which are limited by the knowledge of the ionospheric state, will require a better ability to measure and predict the state of the ionosphere. In particular, global networks measuring the bottom-side electron density profiles over the entire planet would bring major advancements.

There are several possible ways to realize observational systems that would address these issues, each of which has its own issues: A GNSS ground-based and remote occultation total electron content (TEC) network may not have the spatial and temporal resolution needed in propagation applications and data assimilation models. A network of standard and low-power HF sounders across the globe might be feasible, but it is not clear whether oblique networks would work. Oceanographic HF radars operate in the right frequency bands, but to be routinely usable would require standardization of waveform parameters and a network of ionospheric monitoring links to be established using passive receivers. In summary, Ginet said, the HF community needs better bottom-side electron density profiles and parameters over the entire planet, maps of sporadic E-layers, and validated models using operationally relevant metrics (e.g., path loss, group delay, angle-of-arrival). The needs for improved predictive capability include systematic validation of ionospheric models and establishing data integration centers where this validation can occur.

Snively discussed the effects of lower thermospheric dynamics on the bottom of the ionosphere. Atmospheric inputs include thermospheric density and compositional fluctuations as well as strong and variable shears over a wide range of scales. The ionospheric E- and F-regions can be modulated by lower thermosphere neutral waves as well as by acoustic gravity waves, which may include neutral fluctuations. Snively also pointed out that the TEC and HF measurements are valuable inputs not only for ionospheric but also for thermospheric dynamics. Multi-scale, multi-system, physics-based models of thermosphere–ionosphere dynamics in both small scales and across the scales and systems as well as continuous measurements of the most relevant parameters are required for progress in physical understanding and in predictions.

Snively summarized that key observations include high-resolution neutral composition and winds, particularly in the lower thermosphere 100-200 km, and quantification of dynamic processes in the thermosphere that modulate the bottom-side ionosphere. Physics-based modeling of ion-neutral coupling should continue and connect to impacts on signal propagation. Models at different scales should mesh well, be data-constrained, and leverage high-resolution data sets, which will require big-data handling strategies for heterogeneous data sets where the coverage is distributed over space and time.

Bruinsma addressed the large-scale state of the thermosphere–ionosphere system. Thermospheric density is a key parameter for satellite orbit analysis as it is directly proportional to the atmospheric drag force causing spacecraft orbital decay. Currently, the accuracy of orbit calculations is limited by the performance of the thermospheric specification models, which rely on simple empirical models of the atmospheric density. The models are limited by the use of proxies (such as the F10.7 flux) instead of using the actual extreme ultraviolet flux (EUV; see also the discussion of continuous measurements of the solar irradiance in the EUV), by the lack of an accurate description of the neutral composition, and by the inconsistent quality and sparse distribution of data. Keeping track of the increasing number of low Earth orbiting satellites will require precision orbit determination and conjunction analysis, satellite lifetime estimation, and mission analysis.

Improving predictive capabilities in the lower thermosphere–ionosphere, in particular for atmospheric drag calculations, will require improved models in the transition region at altitudes between 100 and 200 km that can capture the net energy input from the solar EUV, the solar wind and interplanetary magnetic field, and the magnetosphere, ionosphere, and thermosphere at sufficiently high temporal cadence.

Bruinsma summarized that observations of composition and density with calibration and processing standards, of sustained measurements in the EUV (of the He II) line, and of solar observations made from the L1 and L5 Lagrange points will be needed. Data assimilation testing schemes, ensemble modeling, and systematic assessment of models (e.g., under the Community Coordinated Modeling Center [CCMC]) will continue to be needed. Lastly, Bruinsma pointed out how important investments in people, workforce expertise and education will be.

Modeling Needs

Another panel addressed observation and modeling needs for the ionosphere–thermosphere system. Its six panelists were Naomi Maruyama of the University of Colorado, Matt Zettergren of Embry-Riddle Aeronautical University, Katrina Bossert of Arizona State University, Bill Lotko of the University Corporation for Aeronautical Research, Larisa Goncharenko of the Massachusetts Institute of Technology, and Hanli Liu of the University Corporation for Aeronautical Research. The panelists were asked to address these key questions:

- What are the advances in modeling and observations needed to improve the understanding of the Sun–Earth system that generates space weather?
- What are the biggest challenges in ionosphere–thermosphere–mesosphere science in the coming decades?
- What observations are needed to test our understanding and our ability to nowcast/forecast the system?
- To what extent is space weather in the ionosphere–thermosphere driven from below?

An overarching theme from this panel was that the whole system—the ionosphere coupled from above and from below—needs to be understood for predicting space weather. The ionosphere between the neutral atmosphere and the magnetosphere is forced from below and from above, and the relative contributions of this forcing depend on location and geomagnetic activity as well as the past history of forcing ("preconditioning"). At high latitudes and during large disturbances, forcing from above dominates, while at middle and low latitudes forcing from below is dominant most of the time. It is as yet uncertain how seasonal and solar cycle variations or climate change shift the relative importance of forcing from below and above or how the upper atmosphere affects the geo-effectiveness of external forcing.

A number of panelists made the point that although multi-scale and multi-step physical processes are important in understanding the ITM system and space weather, they are not well observed or well modeled. ITM variability contains significant energy that isn't captured by the mean state of the system: the large-scale background state can drive small-scale irregularities, and large tides can trigger smaller waves (especially in the polar regions). On the other hand, small-scale turbulence can lead to large-scale changes. Multi-satellite observations are required to resolve the evolution of different waves and their impacts in the ionosphere–thermosphere system.

The single most important factor regulating the magnetosphere–ionosphere interaction is the ionospheric conductivity tying the neutral atmosphere to the ionospheric electrodynamics. The conductivity magnitude and spatial distribution need to be measured at mesoscales (10-100 km) and as a function of time in order to improve the predictive, global geospace models. While it is straightforward to get the conductivity for a given thermospheric state, estimating the precipitation-induced conductivity is more complicated. Since measuring the global conductivity distribution is challenging, the panel suggested combining space-based and ground-based multispectral imaging covering both the global ionosphere and mesoscales down to 10 km resolution.

Ionospheric outflows are a key process through which the ionosphere influences the magnetosphere: They affect magnetic reconnection rates, global dynamics, tail and substorm dynamics, and the ring current and the radiation belts. The observing solutions proposed by the panel included multipoint measurements along flux tubes, global ultraviolet (UV) imaging to determine the ion outflow source and fluxes, and using the upcoming Geospace Dynamics Constellation (GDC) measurements, which, unfortunately, are not optimized for that particular science and will provide only some information. The discussion following the panel presentations touched on imaging solutions, recognizing that some species, such as H^+, cannot be imaged optically, while the imaging of others may present significant challenges.

Gravity waves and tides are another means by which the atmosphere affects the variability of the ionosphere and the plasmasphere. Atmospheric waves are an important source of energy flux from the lower atmosphere and encompass structures in different spatial and temporal scales, such as TIDs, wave dissipation, the Madden-Julian oscillation, and the quasi-biennial oscillation. Understanding how gravity waves affect the circulation and compositional structure of the thermosphere, in terms of both day-to-day variability and smaller-scale waves, requires altitude profile measurements of wind, temperature, and density at a resolution capable of following the evolution of waves with altitude and latitude/longitude. These processes are important for the models, as the thermosphere is highly sensitive to the meridional system, and small changes in thermospheric composition or temperature changes can lead to major changes higher up. Furthermore, climatological measurements of thermospheric parameters (e.g., tides and winds) are required to understand the multi-scale interactions in the ionosphere–thermosphere system. Finally, global-scale thermosphere measurements and models are important in understanding the interaction of forces from below and above and their role in modulating the ITM system during quiet and storm times. Open questions include whether auroral-generated gravity waves produce a different impact than other gravity waves, whether the system predictability changes during storms, and how the lower atmosphere processes affect the fidelity of whole-atmosphere models.

Electric fields and currents connect Earth's atmosphere with the magnetosphere through the ionosphere (i.e., R1 and R2 field-aligned currents) at a range of temporal and spatial scales. The energy injected into the atmosphere as Joule heating is important (as was demonstrated, for instance, by the failure of nearly 40 SpaceX satellites after a geomagnetic storm), but we lack both understanding and models of these nonlinear connections over the full range of temporal and spatial scales. However, we do know that pre-conditioning of the upper atmosphere affects the geoeffectiveness of external forcing; for example, neutral winds increase the variability of the penetration of electric fields. An observational solution would be to monitor the electric currents, as with the approach pioneered by the Active Magnetosphere and Planetary Electrodynamics Response Experiment (AMPERE) and the European Space Agency's Swarm satellite constellation; more spacecraft would improve the spatial coverage and temporal resolution.

A common theme from the panel was that forcing from below is an important aspect of space weather from the plasmasphere out to the magnetosphere. Quantifying this forcing requires high-resolution, four-dimensional observations of the thermospheric parameters (e.g., temperature, O, N_2, O_2 density, vector winds); a further need is to understand the biases between past and current observations and between observations using different techniques. These measurements should also be coordinated with Earth sciences to uncover how climate change may affect the ionosphere–thermosphere coupling (e.g., through increasing temperatures producing convective instabilities [e.g., cyclones, thunderstorms, etc.]), which launch gravity waves, which in turn lead to changes in the upper atmosphere tides and gravity waves.

A new modeling approach to multi-scale physics-based models is needed that can describe that coupling and provide feedback on cross-scale resolution (Figure 3-4). The development of such models will require sustainable support for the research, computational infrastructures, a dense observation network for model validation, and collaboration between the science community and private companies.

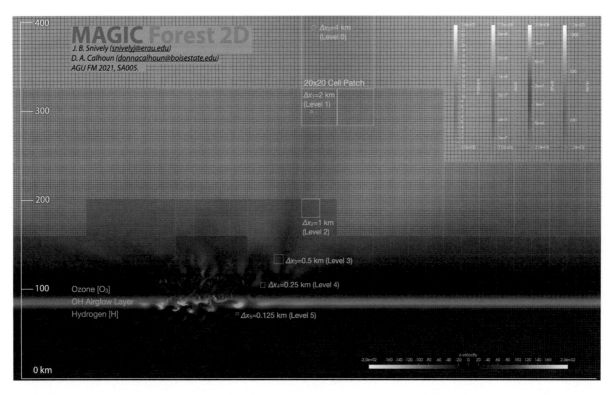

FIGURE 3-4 Recent progress in the numerics and underlying model physics of the Model for Acoustic-Gravity wave Interactions and Coupling (MAGIC) (see Zettergren and Snively, JGR, 120(9), 2015, and references therein) has revolutionized our ability to explore smaller scale phenomena in the ITM. MAGIC is now implemented in ForestClaw (Calhoun and Burstedde, arXiv:1703.03116, 2017), which is an AMR and solver library, using p4est mesh management (Burstedde et al., SIAM JSC, 33(3), 2011), and Clawpack solvers (Clawpack Development Team, 2002-2020; LeVeque, JCP, 131, 1997). This model enables surface-to-space (0-400+ km) acoustic-gravity wave (AGW) dynamics modeling in large (thousands of kilometers) domains, with calculation of species densities and airglow observables. Recent progress has demonstrated applications in long-range, multi-scale AGW propagation and coupling across deep altitude spans and extensions to the ionosphere.
SOURCES: Matt Zettergren, Embry-Riddle Aeronautical University, presentation to workshop, April 12, 2022; from J.B. Snively, "Scalable Modeling of Acoustic-Gravity Wave Interactions, Coupling, and Observables from Surface to Space," abstract SA45A-2202, 2021 Fall Meeting of the American Geophysical Union, December 16, https://agu.confex.com/agu/fm21/meetingapp.cgi/Paper/1001179.

Progress will require sustained investment in local and general-purpose open-source codes that address issues in the current licensing practices. Of specific techniques, it was noted that adaptive mesh refinement (AMR) schemes may be underused. Regarding system predictability, Liu commented that it does not matter "where you put your butterfly"; global data assimilation will be required to manage the error growth.

CROSS-SCALE AND CROSS-REGION COUPLING

Committee member Endawoke Yizengaw moderated a panel on cross-scale and cross-region coupling. The panelists were Josh Semeter of Boston University; Astrid Maute of the University Corporation for Atmospheric Research; Joe Huba of Syntek Technologies; Seth Claudepierre of the University of California, Los Angeles; and Jonathan Rae of Northumbria University in Newcastle upon Tyne, England. The panelists were asked to address the following questions:

- What do we need to understand about cross-scale and cross-region coupling to enable predictive capability of the state of the magnetosphere/ionosphere–thermosphere–mesosphere?
- What are the research needs required to make progress on that understanding?

Semeter began by proposing definitions for cross-scale and cross-region coupling. *Cross-scale coupling* occurs when dynamics at one spatio-temporal scale define boundary conditions for another scale governed by different physics (e.g., processes coupling MHD and kinetic scales). *Cross-region coupling* takes place between geophysical domains demarcated by a change in physical description (e.g., between solar wind and magnetosphere or ring current and plasmasphere). Examples of cross-scale coupling from an energy perspective (see Table 3-1) include tail reconnection at substorm onset driven by ion scale physics

TABLE 3-1 Examples of Cross-Scale Coupling: Energy Perspective

Mode of Energy Transport	Cross-Scale Boundary	Change in Physics	Significance
1. Plasma sheet injection, energization, and dissipation	Mesoscale dynamics (1 Re × 1 minute)	Global MHD → "kinetic" MHD (embedded test particle or PIC simulation).	Ring current dynamics poorly understood. Role of localized injections, waves, and instabilities on storm development not known.
2. Substorm onset	Plasma sheet thickness ~ Ion Larmour radius	Tail reconnection, particle acceleration, breakdown of field theory.	Major mode of internal energy release in the geospace system.
3. Alfvén waves	Wavelength < electron inertial length	Change in dispersion relation, MHD → kinetic.	Dominant mode of energy transport. Dispersion leads to 100-meter variability in precipitation, ionization, and conductivity.
4. Joule heating	Polarization E-fields caused by sub-grid variability in conductance	E-field variability that is either unresolved by the physics model, or undersampled by observations.	Misrepresentation of Joule heating rate. Both over-estimation and under-estimation possible.
5. E-region turbulence (SAID, electrojets, arc boundaries)	$E > 50$ mV/m	Destabilization of modified two-stream instability → nonlinear currents, anomalous heating.	Alters macroscopic conductance seen by magnetosphere.
6. Equatorial plasma bubbles	Irregularity scale near Ion Larmour radius	Non-Maxwellian distributions, dissipation through kinetic instabilities.	Broad-band RF scintillation.
7. RF wave propagation through disturbed ionosphere	Ionospheric irregularities < Fresnel scale (~300 m)	Ionosphere transitions from refractive to diffractive medium.	Stochastic amplitude modulation ("scintillation") and rapid phase fluctuation (loss-of-lock).

NOTE: MHD = magnetohydrodynamics; PIC = particle in cell; RF = radio frequency.
SOURCE: Josh Semeter, Boston University, presentation to workshop, April 12, 2022.

and leading to global magnetotail reconfiguration and energy release; or Alfvén waves transferring energy between the magnetosphere and the ITM system. Many such cross-scale or cross-regional coupling processes are not yet adequately included in the predictive models.

Recognizing the challenge of sampling the geospace system with sufficient resolution, Semeter outlined a scheme to use the ionospheric "projection screen" to deduce information about outer space processes. This would be accomplished by combining ground-based network data to search for cross-scale and cross-region dynamics, which would involve data fusion and inversion models as well as new information theory and AI/machine learning methods and collaboration with computer scientists, engineers, statisticians, and data scientists.

Maute focused on mesoscale electric fields and precipitation (10s to 100s of kilometers) that are critical for accurate estimations of the Joule heating and precipitation; lacking the smaller scales can lead to underestimations by up to 50 percent during active times. Furthermore, the F-region neutral winds respond within minutes to auroral forcing down to 100-km scales and can reduce the local Joule heating. As future needs, Maute identified regionally self-consistent observations of ion drift, particle precipitation, field-aligned currents (FACs), and E- and F-region neutral winds, which are tied to a self-consistent picture by empirical models and data assimilation. On a larger scale, she continued, ionospheric electrodynamics and the geomagnetic fields must be coupled at appropriate scales across physical domains.

Huba discussed cross-scale coupling and irregularities in the ITM using an example of a hierarchy of irregularities in the ionosphere (Figure 3-5) ranging spatially from tens of centimeters to hundreds of kilometers and temporally from milliseconds to hours. This vast range of scales requires the use of different physical equations, from kinetic to fluid theory.

Huba explained that while the majority of current electrodynamics models assume equipotential magnetic field lines, recent work has extended the description to three-dimensional electrodynamics with varying potential along the magnetic field, which allows new instabilities to develop in the models. The community still lacks self-consistent electrodynamic models that would combine the different processes of the low- and high-latitude regions. Such global models would be particularly important for storm times, capturing both the storm time penetration electric fields imposed on the ionosphere by the magnetosphere and the ionospheric storm-time dynamo electric fields. Making progress, he said, will require including scale sizes down to 100s of meters, sub-gridding for small-scale physics, and embedding particle-in-cell (PIC) codes in the global MHD codes.

Claudepierre identified three open questions that need to be addressed to improve the predictive capability of the magnetosphere/ITM system: the role that mesoscale injections of plasma sheet particles play in inner magnetosphere energetic particle dynamics; the significance of energetic particle precipitation (EPP) as a facilitator of magnetosphere–ionosphere coupling; and an improved understanding of global scale cold plasma evolution, energization, and motion.

The ring current formation is an example of the first issue: Is it built up through enhanced global convection, through a series of localized injections, or some combination of both? Estimation of the efficiency of these processes requires a number of assumptions that are not well constrained by either observations or models. This knowledge gap limits the current ability to predict ring current evolution and, thus, the development of geomagnetic storms.

Claudepierre made the following observations as part of a broad strategy to close knowledge gaps: Coordinated multipoint observations distributed widely over the ITM system should include measurements of the injections over a broad region of the nightside plasma sheet. These should be combined with global imaging of the nightside region and concurrent measurements of the ring current and radiation belt particles. In addition, these measurements need to be coordinated with a robust modeling program, such as MHD test-particle simulations or global MHD with embedded PIC simulations. A particular need is

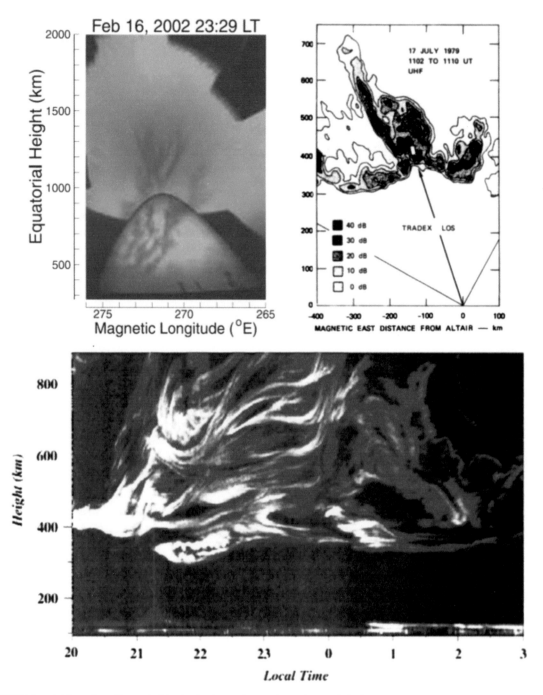

FIGURE 3-5 A hierarchy of irregularities in the ionosphere where the top left shows optical emissions observed from Mount Haleakala showing irregularities on the scale of tens of kilometers, the top right shows the 11-cm radar backscatter from Kwajalein Atoll in the Marshall Islands, and the bottom shows the 3-m radar backscatter from the Jicamarca Radio Observatory in Peru. SOURCES: Joe Huba, Syntek Technologies, Inc., presentation to workshop, April 12, 2022. Top left: M.C. Kelley, J.J. Makela, B.M. Ledvina, and P.M. Kintner, 2002, Observations of equatorial spread-F from Haleakala, Hawaii, *Geophysical Research Letters* 29(20):64-1–64-4, Copyright 2002 by the American Geophysical Union; Top right: Defense Nuclear Agency, 1979, "Backscatter Measurements of 11-cm Equatorial Spread-F Irregularities," Washington, DC: Department of Defense, https://apps.dtic.mil/sti/pdfs/ADA091980.pdf; Bottom: Courtesy of David Hysell, Cornell University.

to improve the accuracy of global geomagnetic field models, as the mapping from the magnetosphere to the ionosphere plays a crucial role in the coupling studies. Concerning implementation, Claudepierre said that the HSO has not been strategically planned from a systems science perspective and that multipoint measurements could be realized by adding sensors onboard operational missions such as the polar-orbiting environmental satellites (such as the Polar Operational Environmental Satellites [POES] system) or the Defense Meteorological Satellite Program (DMSP), which have traditionally provided such measurements. It will be important both to ensure the long-term continuity of such data and to establish new programs.

Rae began by noting that cross-scale and cross-region coupling in space plasmas covers a vast range of spatial and temporal scales as well as of plasma characteristics present throughout the heliosphere, from the solar corona to the ionosphere. In particular, he said, it is not possible for any single simulation to cover all of the associated energy ranges.

Solar wind energy input can only account for 50-60 percent of the magnetic perturbations in the ionosphere, with the other 40-50 percent likely coming from the magnetosphere associated with substorm activity. This second component is important, as substorms power the radiation belts and the ring current and deposit energy into the ionosphere and drive geomagnetically induced currents, which are a space weather hazard. A further problem with predicting magnetosphere–ionosphere coupling and space weather impacts, Rae said, is the lack of understanding of the radiation belts. It is important to understand how the magnetotail feeds the radiation belts, including cold plasma (i.e., on the order of eV), reconnection, substorms, wave propagation, and wave–particle interactions; warm plasma (keV) for wave–particle interactions, wave propagation, and spacecraft charging; and relativistic plasma (MeV) for single-event upsets. Multi-point measurements are needed to verify the data-driven models.

For the future, new point measurements in the key regions of geospace will be needed to both drive and test models, with continuous ground-based measurements providing key global measurements. Realistic models of plasma behavior for substorms, cold plasma, warm plasma, and relativistic plasma are needed; they would build on these new data. And it will be important to determine what can be predicted deterministically and what is best described by probabilistic models.

REFERENCES

Adewuyi, M., A.M. Keesee, T. Nishimura, C. Gabrielse, and R.M. Katus. 2021. "Mesoscale Features in the Global Geospace Response to the March 12, 2012 Storm." *Frontiers in Astronomy and Space Sciences* October 27. https://doi.org/10.3389/fspas.2021.746459.

Blum, L.W., L. Kepko, D. Turner, C. Gabrielse, A. Jaynes, S. Kanekal, Q. Schiller, et al. 2020. "The GTOSat CubeSat: Scientific Objectives and Instrumentation," P. 113892E in *Micro-and Nanotechnology Sensors, Systems, and Applications XII*, Vol. 11389. Bellingham, WA: International Society for Optics and Photonics.

Claudepierre, S.G., and T.P. O'Brien. 2020. "Specifying High-Altitude Electrons Using Low-Altitude Leo Systems: The SHELLS Model." *Space Weather* 18(3):e2019SW002402. https://agupubs.onlinelibrary.wiley.com/doi/full/10.1029/2019SW002402.

Delzanno, G.L., J.E. Borovsky, N. Buzulukova, C.R. Chappell, M. Denton, P. Fernandes, R. Friedel, et al. 2021. "The Need to Understand the Cold-Ion and Cold-Electron Populations of the Earth's Magnetosphere: Their Origin, Their Controlling Factors, and Their Impact on the System." *Heliospheric 2050 White Papers*, 4033. https://www.hou.usra.edu/meetings/helio2050/pdf/4033.pdf.

Fernandes, P.A., G.L Delzanno, M.H. Denton, M.G. Henderson, V.K. Jordanova, T.K. Kim, B.A. Larsen, et al. 2021. "Heavy Ions: Tracers and Drivers of Solar Wind/Ionosphere/Magnetosphere Coupling." *Heliospheric 2050 White Papers*, 4047. https://www.hou.usra.edu/meetings/helio2050/pdf/4047.pdf.

Goldstein, J., D.L. Gallagher, P. Molyneux, and G.D. Reeves. 2021. "Core–Plasma Refilling and Erosion: Scientific Justification." *Heliospheric 2050 White Papers*, 4063. https://www.hou.usra.edu/meetings/helio2050/pdf/4063.pdf.

Ilie, R. 2021. "The Need for Detailed Ionic Composition." *Heliospheric 2050 White Papers*, 4094. https://www.hou.usra.edu/meetings/helio2050/pdf/4094.pdf.

4

Research, Observation, and Modeling Needs: Ground Effects

Key Themes

In the sessions devoted to research, observation, and modeling needs, the following key themes emerged from the presentations and discussions:

- In all areas of space weather, dealing with the entire system of systems is important to being able to understand and predict the system's behavior.
- Given the importance of multiple viewpoints and global coverage, inter-calibration of the observing instruments is crucial, as are robust methods for incorporating data into the models.
- More observations covering all aspects of extreme events are needed to understand their causes, variability, propagation and evolution, and impact on individual infrastructures, such as power grids.
- A dense network of ground-based geophysical measurements, including those provided by the private sector, will be required to better understand and provide high-confidence, long lead-time predictions of geomagnetic disturbances and geomagnetically induced currents.

Two workshop sessions were focused on research needs and on observation and modeling needs related to ground effects. The first of these was a keynote presentation on research needs for the prediction of ground effects, and the second was a panel devoted to the observation and modeling needs for ground effects. As previously noted, these presentations specifically addressed three different parts of the statement of task:

- *Examine trends in available and anticipated observations, including the use of constellations of small satellites, hosted payloads, ground-based systems, international collaborations, and data buys, that are likely to drive future space weather architectures; review existing and developing technologies for both research and observations.*
- *Consider the adequacy and uses of existing relevant programs across the agencies, including NASA's Living With a Star (LWS) program and its Space Weather Science Application initiative, the National*

Science Foundation's (NSF's) Geospace research programs, and NOAA's Research to Operations (R2O) and Operations to Research (O2R) programs for reaching the goals described above.

- *Consider how to incorporate data from NASA missions that are "one-off" or otherwise non-operational into operational environments, and assess the value and need for real-time data (for example, by providing "beacons" on NASA research missions) to improve forecasting models.*

The following two sections summarize those ground effects–related presentations.

RESEARCH NEEDS

In a keynote presentation on the workshop's first day, Jeffrey Love of the U.S. Geological Survey (USGS) spoke about the geoelectric hazards and effects on U.S. power grids caused by geomagnetic storms as well as about what is needed for better prediction of such storms.

Geomagnetically induced currents (GICs) are among the most important hazards of space weather, he said, because they can give rise to severe, irreversible damage on power grids. The GICs are driven by induced geoelectric fields, which depend on both the intensity of the geomagnetic disturbance (ionospheric currents) and the Earth-surface impedance (subsurface electrical conductivity structure determined by geology). The overall magnitude of the risk is controlled by the externally driven storm intensity.

The Carrington event in 1859 is often used as a reference for a superstorm, but in the twentieth century alone, five significant geomagnetic storms took place. One of these, which struck in March 1989, caused significant damage to, among many other things, a high-voltage transformer at a nuclear power center in Salem, New Jersey (Figure 4-1). At a 2008 National Academies workshop, participants predicted that a superstorm similar to the Carrington event could cause significant damage and interference to military and civilian assets, including the possibility of widespread and prolonged loss of electricity, damage to the grid, and severe disruptions to Global Positioning System, radio communication, and geophysical surveys.[1] Estimates provided at the workshop were that the total economic impact just for the United States would be between $1 trillion and $2 trillion. Although such an event has not been experienced since the development of modern technology, history makes it clear that a future occurrence is plausible.

The geographical distribution of geoelectric hazards is complex, as the surface electromagnetic impedance can differ significantly from one location to another depending on subsurface mineralogy and fluid content. A geoelectric hazard analysis combined with the anomalies record associated with the March 1989 geomagnetic storm indicated that if an even more intense storm were to strike the mid-Atlantic and northeastern United States today, it would likely lead to much more serious problems than those seen after that earlier storm. The USGS has a goal to establish a more complete geomagnetic monitoring and to conduct a dense wideband magnetotelluric survey of the mid-Atlantic and northeastern United States.

The impacts of the storm-time geomagnetic disturbance also depend on the details of the power grid design. Therefore, to predict and prepare for future geomagnetic storms, additional geomagnetic monitoring is needed, especially focusing on event-by-event mapping of voltages along individual grid lines. Furthermore, the E3 component of a nuclear electromagnetic pulse (EMP) following a nuclear weapon detonation can cause GICs like those generated during geomagnetic storms, highlighting the need to continue the hazard analysis for cross-cutting research.

---

[1] National Research Council, 2008, *Severe Space Weather Events: Understanding Societal and Economic Impacts: A Workshop Report*, Washington, DC: The National Academies Press, https://doi.org/10.17226/12507.

March 1989 magnetic storm damage to a high-voltage transformer at a nuclear power center in Salem, New Jersey.

FIGURE 4-1 Damage to a high-voltage transformer caused by a geomagnetic storm.
SOURCES: Jeff Love, U.S. Geological Survey, presentation to workshop, April 11, 2022; courtesy of Public Service Electric & Gas (PSE&G) Company New Jersey.

OBSERVATION AND MODELING NEEDS

A panel moderated by committee member Delores Knipp was devoted to the observation and model needs related to ground effects. The panelists were Adam Schultz of Oregon State University, Jenn Gannon of Computational Physics Inc., Jesper Gjerloev of the Applied Physics Laboratory of Johns Hopkins University, Arnaud Chulliat of the University of Colorado and NOAA's National Centers for Environmental Information, Antti Pulkkinen of NASA's Goddard Space Flight Center, and Anna Kelbert of USGS.

The focus of this panel was on geomagnetic disturbances (GMDs) and their associated ground effects and GICs. The panelists' presentations had three overarching themes.

First, more information is needed on extreme space weather conditions: the ranges of possible impacts are not well understood, and there is a lack of understanding of their causes. Second, global simulations, other models, and the combination of observations with models are crucial for developing user-friendly and comprehensive space weather products—one wiggly line from an instrument is not sufficient for either understanding or predicting space weather. It is important to implement real-time data assimilation into global models (a theme that also emerged from the Data Assimilation panel) as well as to develop higher-level data products for the user community. Experience has shown that community-wide validation activities are crucial for taking models from research to operations. Third, the scarcity of global and continuous real-time measurements is a particularly crucial hindrance of space weather forecasting capabilities. A dense network of geophysical observatories, including ones provided or operated by the private sector, will be required to better understand and provide high-confidence, long-lead-time predictions of GMDs and GICs (see also New Architectures panel discussion in Chapter 6). Continuous maintenance of the ground-based observatories hosting magnetometers, all-sky imagers, or Fabry Perot interferometers requires long-term funding support.

The need for a dense network of ionospheric observations arises from the high spatial and temporal structuring of the aurora and auroral electrojet currents. A panelist gave an example of an intense, localized magnetic field decrease of over 2,200 nT during only 9 minutes recorded at a station in Iceland, which was not observed at all at the surrounding magnetic stations. Furthermore, predicting the GICs driven by the auroral electrojet currents will also require a dense network of observatories that provide high-resolution, three-dimensional conductivity models. As an example, Denver, Colorado, lies within a localized hazard region due to a geological boundary that would not show up in a low-resolution conductivity model (see Figure 4-2). A denser magnetotelluric survey conducted by a denser geomagnetic monitoring network

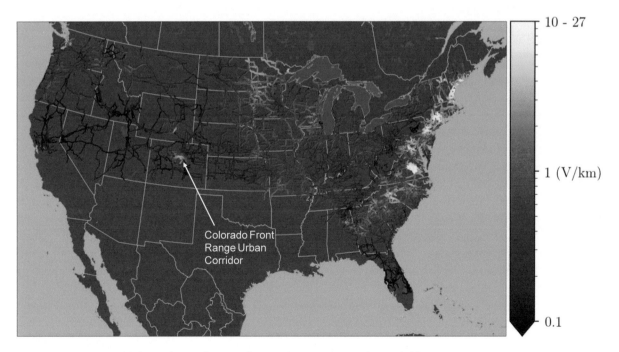

FIGURE 4-2 Nation-wide hazard map showing the 100-year maximum electric field.
NOTES: Lighter regions (higher electric field) indicate a higher risk. Later data ruled out the localized hazard region around Denver as an artifact of the data.
SOURCES: Anna Kelbert, U.S. Geological Survey, presentation to workshop, April 12, 2022; from G. Lucas, J.J. Love, A. Kelbert, P.A. Bedrosian, and E.J. Rigler, 2020, "A 100-Year Geoelectric Hazard Analysis for the U.S. High-Voltage Power Grid," *Space Weather* 18:e2019SW002329, https://doi.org/10.1029/2019SW002329, © 2019 The Authors.

is critical for creating a high-resolution ground conductivity map, which can be turned into a nationwide hazard map. The last step in observations is to have an up-to-date power grid system configuration, which allows computation of the actual GICs in the system.

Additionally, the re-establishment of active connections between scientific and engineering communities will be required to address issues related to GIC modeling and analysis. The North American Electric Reliability Corporation (NERC) GMD task force, one of the main forums attempting to combine and coordinate effort between scientists and engineers for this purpose, was recently reorganized under the Reliability and Security Technical Committee.[2] The link between engineers and scientists is crucial for providing rigorous validation of the magnetotelluric data used in GIC applications.

There are many possible ways to create a dense observational network: For data distribution, the SuperMAG international collaboration of ground-based magnetometers hosted at the Applied Physics Laboratory of Johns Hopkins University is a good example of an implementation strategy. For data production, one path forward could be to develop collaborations with the private sector to share the cost of instrument deployment and operations. The private sector in need of space weather information has underused potential to support collection of the relevant information. For example, one group of small businesses provided more than 85 full-time equivalents of support to federal government space weather efforts.

Possible other data sources include those provided by the IRIS (In Situ and Remote Ionospheric Sensing) suite, which may provide a partial solution to developing estimates of the ground effect over the United States. In the future, the Electrojet Zeeman Imaging Explorer (EZIE) mission will map the structure and evolution of the auroral electrojet by flying three CubeSats in a string-of-pearls configuration and making Zeeman splitting measurements.

Another proposed observing solution would entail combining highly accurate, observatory-level, full-field measurements and networks of science-grade variometers with deployment of tens of thousands of amateur-grade variometers. Data from this network combined with observations from space missions and assimilated into a model could provide real-time space weather nowcasts and forecasts. For example, the outreach effort of EZIE will deploy 1,000 magnetometers across the United States, which, while not science-grade, will be sufficient for space weather (science) purposes. This effort illustrates one way to deploy 10,000 magnetometers on a reasonable budget.

Models of the internal geomagnetic field—such as the World Magnetic Model (WMM) and the International Geomagnetic Reference Field (IGRF)—are important, while ionospheric and magnetospheric currents, especially during storms, cause perturbations to the main field. The internal field models have various applications, including orientation and navigation (for airplanes, ships, submarines, spacecraft attitude control, antenna tracking), directional drilling (energy industry), and alternative positioning and navigation (MagNav). Currently, the WMM is specified based on data from low-Earth-orbit satellites (such as the European Space Agency's SWARM satellite constellation and the Iridium satellite constellation), ground observatories (the international INTERMAGNET network of ground magnetic observatories), and information on the solar wind conditions from L1 (the Deep Space Climate Observatory, or DSCOVR). However, despite the various sources of data, observational gaps remain that limit the accuracy of the geomagnetic field models.

---

[2] North American Electric Reliability Corporation, "Reliability and Security Technical Committee (RSTC)," https://www.nerc.com/comm/RSTC/Pages/default.aspx, accessed August 10, 2022.

5

Modeling, Validation, and Data Science

Key Themes

In the sessions devoted to modeling, validation, and related topics in data science and analytics, the following key themes emerged from the presentations and discussions:

- The space weather field is simultaneously data starved and not efficiently using available data.
  - Existing data are often not well matched to current modeling needs because of the data format, latency, lack of metadata, lack of calibration or intercalibration, or limited knowledge of data errors. In many cases, models need additional data that are not currently available.
- Data assimilation (DA) is a tool that holds significant promise for space weather research and forecasting. DA is not a "one-size-fits-all" approach, rather its use will require tuning across domains that cover disparate physics and vast ranges of temporal and spatial scales.
  - The current potential for applying DA to space weather models and predictions is limited by a lack of (1) suitable data; (2) characterization of data uncertainty; and (3) computational capacity. Commercial data buys may be part of the data solution; however, the error sources in such data will need to be well documented. Cloud computing capacity is growing but may turn out to be an expensive solution.
- Ensemble modeling, particularly with multi-model ensembles, has the potential to significantly improve space weather predictions; however, the widespread use of ensemble models will require additional research and development as well as additional resources, such as greatly increased data storage and computational capacities.
- Machine learning (ML) is a promising approach for understanding space weather, but its use is currently limited by data quality and data quantity. Furthermore, the space weather community does not yet have the data ecosystem needed to create ML-ready data sets. All domains lack data related to extreme events. In ML space weather applications, a greater emphasis is needed for understanding the underlying physical phenomena, as opposed to typical ML applications treating the model as a "black box" and the prediction as the only final outcome.
- Robust uncertainty quantification methodologies will be important for approaching space weather as a system science. Model uncertainty and data-representativeness uncertainty will need to be quantified in a systematic way across the different models.
- Observing system simulation experiments (OSSEs) can be valuable in cost–benefit evaluations of particular data sets to be used in ML and DA approaches, but not all domains are mature enough for the OSSEs to be useful.
- Space weather data science would benefit from further cross-agency effort to coordinate data-archival standards, promote data fusion and reuse, and support data revitalization for ML, ensemble modeling, and data assimilation.

- Resources are needed to archive and curate data and model outputs in such a way that the findable, accessible, interoperable, reusable (FAIR) principles are satisfied.
- Advancing space weather data science will require investment in the workforce.

The committee's statement of task requested it to "consider needs, gaps, and opportunities in space weather modeling and validation, including a review of the status of data assimilation and ensemble approaches." These topics were addressed in the workshop's Data Science and Analytics session, which was spread over days 3 and 4 of the workshop. This chapter summarizes the contents of two keynote talks and four panels from that session, which addressed different aspects of modeling and validation challenges in the context of advances in computational resources, tools, and ongoing needs, especially in terms of data curation.

The background against which the panel's discussions took place is the profound contradiction that exists in space weather research; that is, it is in a "big data" regime, yet not enough data are available for data science applications. This discipline is unique in the need to understand (and compute) relevant physics on scales that span meters to megameters and sub-seconds to decades as well as across vast density, temperature, and electromagnetic field scales. Thus the modeling and data challenges facing space weather scientists are significant.

The 2016 National Science Foundation (NSF) Portfolio Review of the Geospace Section of the Division of Atmospheric and Geospace Science briefly mentioned aspects of data science in the recommendations for data exploitation tied to the previous Heliophysics Decadal Survey. The gap analysis for the National Aeronautics and Space Administration's (NASA's) Space Weather Science Application Program focused on at-risk and needed-but-not-yet-available observations. Within the gap analysis the topics of data science and analytics were not directly addressed, although a few of the sub-elements such as data assimilation (DA) and ensemble modeling were mentioned. Data science and analytics in support of space weather system science was not addressed specifically in either report.

KEYNOTES: DATA ASSIMILATION AND MACHINE LEARNING

The workshop's data science and analytics sessions were kicked off by two keynote speakers, Richard Todling of NASA's Goddard Space Flight Center and Enrico Camporeale of the University of Colorado Boulder and the National Oceanic and Atmospheric Administration's (NOAA's) Space Weather Prediction Center. The keynote speakers were asked to address two questions:

- Compared with "Earth system science," space science is usually data starved. How do we leverage data assimilation and machine learning to overcome this?
- What is the role of testbeds in paving the way for data assimilation and machine learning from research to operations?

From Terrestrial Weather to Space Weather

Todling's presentation began with a brief overview of the use of DA in terrestrial weather forecasting. DA is a mathematical method in which observations and numerical model data are combined to create an optimal representation of the state of a system. In the case of terrestrial weather prediction, a typical DA approach is to compare data from various satellites and ground-based observations with a previous numerical model forecast and then update the model state to better match those new data. This is an iterative process in which model states are constantly being revised to reflect observations of the present state to produce a new forecast for the next time step, which is then used as a new basis of data comparison.

Terrestrial weather prediction has steadily improved because of several factors, Todling said: increased model resolution, improved representation of physical processes, and the increasingly large amounts of data from observations that are assimilated into the models through advanced DA techniques. Over time, a wide variety of different DA techniques have been developed. There are two main DA types—variational DA and sequential DA—each with its own variations, and many of these have been combined with ensemble models to create hybrid DA models.

Similarly, he said, there are a large number of different machine learning (ML) approaches to learning from data without guiding by models. In essence, ML looks for patterns in large collections of data that are used as the basis for making predictions. Compared to DA, in ML there are no a priori models that the data build on or are compared with. There are also a large number of hybrid models in which ML is used to assist DA or DA procedures are used to aid ML strategies (Figure 5-1).

With that background, Todling turned to a discussion of DA and ML in space weather and to comparing that use with their application in terrestrial weather forecasting. DA is used in a large number of space weather models covering the photosphere, the corona, flares, coronal mass ejections and solar wind, the magnetosphere, the ionosphere, and the thermosphere–ionosphere system. One difference between terrestrial and space weather systems, he noted, is that the spatial scales are greater and the particles and

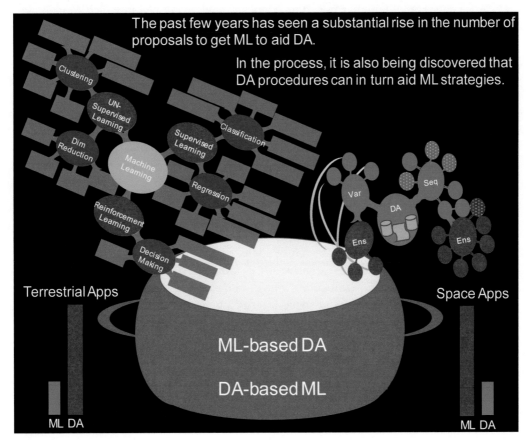

FIGURE 5-1 Combining data assimilation (DA) and machine learning (ML) to tackle challenges of space weather forecasting. In tandem, these two methods provide a large toolkit with which to tackle space weather forecasting challenges.
SOURCE: Ricardo Todling, NASA Goddard Space Flight Center, presentation to workshop, April 13, 2022.

electromagnetic disturbances travel across the system in time scales much faster than in the terrestrial atmosphere. The two types of methods, ML and DA, can play different roles in providing a hybrid tool for forecasting, with DA being used both within a domain and to bridge domains, and ML being used, for example, to provide process emulation, data down-scaling, and homogenization. Among the different DA and ML models used in both terrestrial and space weather forecasting, the models may be intercoupled to a varying degree, depending on the application.

The many different facilities and methods based on DA and ML have been key in terrestrial weather research for bridging temporal and spatial scales of relevance to the specific domain. As the role of uncertainties is becoming more central to evaluating forecasts and to providing actionable predictions, Todling said, it is important to understand that DA and ML do not, by themselves, provide error estimates, but they can be used for example in ensemble or Monte Carlo–type applications, which can provide such estimates.

The terrestrial weather modeling community has adopted a few standard modeling frameworks within which to develop further capabilities, although, according to Todling, that community has largely decided that modeling frameworks are not adequate for application of or research on DA techniques. In the area of space weather forecasting, DA and data-informed "nudging" have been used in modeling across solar, heliosphere, magnetosphere and thermosphere/ionosphere domains but mostly in "proof of concept" studies, not yet reaching an operational state.

The space weather community has begun to establish frameworks within which models—especially cross-domain, cross-system models—can be developed, but it is not clear that there is a need (yet) for a single space-weather-focused DA/ML framework. Frameworks in general allow flexibility via module-based units at each step. The Joint Effort for Data-Assimilation Integration (JEDI) framework has been adopted across some research programs for use in terrestrial weather modeling, Todling said, and it could possibly be appropriate for the space-weather regime as well.

A key message from Todling's presentation was that there is not a one-size-fits-all approach that will work across all physics-based domains or disciplines. The DA experience from terrestrial weather modeling can and should provide lessons learned for application to space weather research, but DA is still a relatively new tool for space weather, and its use in that area can be expected to have challenges requiring unique solutions.

Machine Learning in Space Weather Forecasting

In the next presentation Camporeale began by claiming that ML is "reinventing space weather." To back up this contention he provided an extensive list of topics in space weather to which ML has been applied, including the forecasting of global or average indices, such as the disturbance storm time (Dst) index, solar wind classification, solar wind speed forecasting, and predicting the arrival time of coronal mass ejections (CMEs). The potential to transform space weather, he said, derives from the fact that ML holistically estimates a system's behavior, while physics-based models are deterministic, describing only processes that are included in the governing equations.

In explaining why ML is so well suited to space physics problems, Camporeale said that the basic reason is that physical properties, such as invariance and symmetry, along with conservation laws, drastically reduce the search space of the ML parameters. He added that ML should be able to describe any system that follows the laws of physics. The major challenges to the use of ML in space weather are data quality and data quantity.

In examining the path forward for ML in space weather, Camporeale offered a series of questions and challenges that must be addressed. For example, a pervasive question in research on ML-based space weather forecasts is what he called "the information problem," the challenge of determining the minimum amount of physical information required to make a forecast. The "gray-box problem" centers on the question of the best way to optimally use physics understanding and the large data set covering the Sun–Earth system. For example, in the case of probabilistic estimates of regional ground magnetic perturbations, it was shown that estimates provided both by physical models and by pure ML were inferior to approaches combining the two.

The "surrogate problem" focuses on determining which components in the "Sun-to-mud" chain can be replaced by an approximated black-box surrogate model with an acceptable trade-off between computational speed and decreased accuracy. The "uncertainty problem" involves determining how to incorporate the uncertainties in the data throughout the model outcomes and ML-based forecasts. Camporeale said that propagating uncertainties through the space weather chain from solar images to magnetospheric and ground-based observations to a single-point prediction is a complex and computationally demanding task.

The "too often too quiet problem" arises from the fact that as the geomagnetic storms are rare, the space weather data sets are imbalanced, being dominated by quiet conditions. This creates a serious problem for the ML algorithms, he said, and it also poses challenges for defining meaningful metrics that assess the ability of a model to predict interesting but rare events.

The final challenge on the path forward is the "knowledge discovery and explainability problem," which centers on distilling some understanding of the physical mechanisms from the ML-based black-box predictions.

In closing, Camporeale said that these six problems (the information problem, the gray-box problem, the surrogate problem, the uncertainty problem, the rare events problem, and the knowledge discovery and explainability problem) are not specific to space weather but also pose fundamental challenges in the fields of artificial intelligence (AI) and uncertainty quantification.

Discussion

The discussion following the session addressed a number of questions from the Zoom chat. Many of the two speakers' comments were focused on the possible roles of ML, interpretable ML, training intervals for ML-based models, and the lack of ML-based models in scoreboards. The role of ML as compared with physics-based models was also discussed, with comments such as "ML cannot learn what has not yet [been] seen" and questions such as "Can ML help inform which measurements we will need in the future to improve forecasts?"

Camporeale said that ML can have a clear role in populating a well-delineated "state space" that is otherwise sparsely observed. Furthermore, ML may play a role in identifying less than catastrophic but still very rare disturbances that have major effects on society, such as the Starlink failures that occurred during a time of minor space weather activity.[1] Overall, participants' comments indicated that the useful role of ML processes is not yet fully understood or appreciated.

Finally, there was some discussion as to whether the space weather DA applications are sufficiently similar to those used in terrestrial weather applications to make JEDI a useful tool for space weather.

MACHINE LEARNING AND VALIDATION

The Data Science and Analytics: Machine Learning and Validation panel, moderated by committee member KD Leka, had presentations from Jacob Bortnik of the University of California, Los Angeles; Asti Bhatt of SRI; Shasha Zou of the University of Michigan; Morris Cohen of the Georgia Institute of Technology;

---

[1] T. Malik, 2022, "SpaceX Says a Geomagnetic Storm Just Doomed 40 Starlink Internet Satellites," Space.com, February 9, https://www.space.com/spacex-starlink-satellites-lost-geomagnetic-storm.

David Fouhey of the University of Michigan; and Hannah Marlowe of Amazon Web Services. The panelists were asked to address two questions:

- What investments are needed to produce physics-informed machine learning for space weather?
- Are there adequate data sources and curation to implement machine learning for research, validation, and the determination of uncertainty?

A panel discussion followed the presentations.

Bortnik began the presentations by providing an overview of how ML can be used to extract knowledge from data. In general ML applications the goal is to build a black-box model that can make forecasts and predictions, which do not offer any insights into the underlying physical mechanisms. He said that there should be dual goals of both building a model that predicts well and extracting physical insight and understanding. There are various ways to use ML that allow developing these physical insights, including requiring ML systems to obey physical laws (e.g., ensuring symmetries, invariants, and conserved quantities), using interpretable models (transparency into which inputs control which outputs), using models that show how information gets transmitted in the system (i.e., causality flow), and extracting governing equations from the models to gain understanding of the physics of the system. Bortnik noted that it will be important to incorporate these elements into space weather ML systems, even if such approaches are not yet well developed.

In the future, Bortnik said, all space scientists will need a working knowledge of ML both because the rapidly growing data volumes cannot be analyzed in traditional ways and because ML supersedes physics models in many cases. Thus, space science education will need to include ML principles as well as provide experience in building ML models. This will require new books, classes, and curricula; more workshops and meetings devoted to the topic; and the development of new and more powerful algorithms for analyzing physical systems. Figure 5-2 summarizes ideas for the application of ML in the Earth and space sciences.

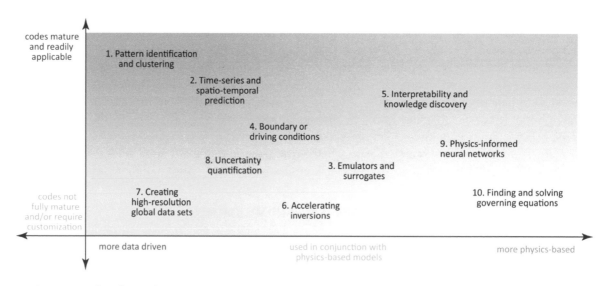

FIGURE 5-2 Ten ideas for applying machine learning in the Earth and space sciences.
SOURCE: Jacob Bortnik, University of California, Los Angeles, presentation to the workshop, April 13, 2022.

Next Bhatt spoke about ML in space science from the perspective of a data provider. She said that while ML methods have been around longer than most large data sets, it was investment in the creation of ML-ready data sets that led to the expansion of their use in ML applications. However, she said that, generally speaking, the space weather community does not yet have the data ecosystem needed to create ML-ready data sets. The space physics data infrastructure, especially the ground-based observations, is severely lacking in its ability to provide standardized data products. As a result, ML today is being applied to only a small subset of available space weather data.

Bhatt argued that space-weather data providers need to adopt standardized metadata as well as the FAIR practices—that is, ensuring that data are findable, accessible, interoperable, and reusable. At this point, most space-weather data do not fulfill those requirements. To ensure that FAIR practices are universally applied to both data and model-output, the space-weather community and the funding agencies need to support the enactment of FAIR data.

In the next presentation, Zou highlighted "melting" boundaries between disciplines and answered the session's two questions with that focus in mind. Concerning the investments needed to produce physics-informed ML for space weather, she emphasized the need for "sustained yet agile funding programs" to support interdisciplinary teams and interdisciplinary academic courses and programs designed to train the next generation of space physicists and, in particular, to equip them with the skills and knowledge they need to apply ML in their research. On the question of adequate data sources for ML applications, Zou said that data are still sparse in large part due to the disparate spatial and temporal scales relevant to space weather phenomena. She suggested establishing satellite constellations that would provide multi-point observations in diverse space plasma regimes. She also said that closer collaboration is needed between government agencies and the private sector to harness the boom of the private space industry and AI technologies.

Morris, considering the question on the adequacy of data, highlighted the fact that while space science data are "big data," they are also misbehaving data, rife with missing and inconsistent information. Space weather ML applications aim to produce interpretable outcomes from a fusion of theory, models, and multiple data sets that span decades, often without overlap for intercalibration. As such, these problems are of interest also to the data science community, and he recommended marketing the potential for new advances applicable also to other research fields to data scientists. To encourage advances in this area, Morris suggested incentivizing space-based data science course development at computing programs and departments, funding new space scientist faculty hires in computer science departments, and specifically tackling the challenge of "misbehaved data" in data science using the space weather data as an example.

Fouhey, an ML and computer vision (CV) expert involved in space weather problems, spoke about how to get more ML and CV people involved in space research. First, it is important to provide easy entry points for ML experts. Second, space weather problems that are interesting and have research benefits and societal impacts should be on offer. Beyond this, Fouhey noted that it takes time to become proficient in a new field; therefore, ML experts would benefit from discipline-scientist-led "welcome mats" (such as the NASA-established frontier development labs) as well as sustained cross-discipline funding. Looking to the future, Fouhey said that nurturing strong collaborations between space-weather research and ML-centered research could lead to (1) data pipelines and instruments that let advances in ML aid space weather research, (2) work that advances both ML and space weather, and (3) the co-design of future instruments optimized for ML-based support and investigations.

Finally, Marlowe offered a high-level view of what is important in a successful use of ML: (1) Make sure that the data are ML-ready—in particular, that they are available and optimized for consumption. (2) "Get technology out of the way" so that researchers can focus on the most important parts of the problem. (3) Make sure that the right people are available through collaborative teams that have a blend of technical and domain experts. (4) Ensure appropriate training and knowledge sharing so that workers can effectively ask and answer questions of when or whether to apply ML. Finally, he concluded, it is important to encourage and lean in to new and innovative approaches and learn from failures across the disciplines.

Discussion

The first topics addressed in the panel discussion were how to engage ML experts in solving space weather questions and, in particular, how to establish collaborations by building on the "low-hanging fruit." One suggestion for an entry point was using ML to classify data in various ways, a process that would require both physical understanding and ML techniques. But there does need to be an acknowledgment that gray-box/black-box approaches can make physicists inherently uncomfortable.

Two questions to the panel were posed by the moderator: Were there sufficient data and curation available to use ML for research, validation, and uncertainty quantification, and how does one go from a curated data set being used to test an ML application to using ML in operations, where the data are not nearly as high quality, continuous, or validated?

Bortnik said that the answer will probably always be "no" to having enough data, but that should not prevent the pursuit of these approaches. Fouhey continued that the handling of operational (lower-quality) data is already a data science research topic, which reinforces the point that space physics provides research topics that are of interest to ML and data scientists as well. Marlowe emphasized the importance of the "unglamorous" but time-consuming data preparation and curation tasks that can take more than half of the time of any ML project. Tomoko Matsuo spoke about the role of observing system simulation experiments (OSSEs) in ML validation in evaluating model performance in different data ecosystems.

Crowd-sourcing through "coding challenges" has been successfully used in many fields to attract broad participation from the public and from new research communities. Jack Ireland noted that these require clear evaluation benchmarks and well-posed questions. Defining evaluation benchmarks for ML space weather applications could help their development as well as assisting in crowd-sourcing and attracting broader awareness (e.g., through prizes) of these topics. Public engagement could be enhanced through prize-offering challenges, such as the recent TopCoder challenge to develop comet-detecting algorithms, for the dual-purpose of advancing space science and bringing the excitement of space physics to a broad group of coding experts.

Lastly, classroom education in geosciences programs is needed at both the undergraduate and graduate level; topics would include ML pre-curated data sets and benchmarks as well as space weather domain knowledge.

Zou made the comment that it might be possible to encourage cross-discipline training by offering various incentives, such as certificate programs for students in other fields. While some certificate programs in ML and statistics exist, there are only a few opportunities for ML and statistics experts to get certificates in space weather subdisciplines.

The session also inspired a spirited discussion on the Zoom chat concurrent with the presentations. Strong arguments were made that all scientists should have ML expertise, and there was strong support for choreographed cross-disciplinary programs (e.g., the Frontier Development Lab developed by NASA/JPL). In developing these collaborations, it will be important to ensure that the problems and ML approaches are interesting to both parties, offering analysis-ready challenging data sets to the ML experts and novel methodologies to the physicists. For example, the Earth Science Information Partners (ESIP) Data Readiness Cluster provides standards for developing AI-ready data sets.

Two questions were posed concerning the potential power of OSSEs: (1) Are the models sufficiently well-defined that the OSSE methodology makes sense? and (2) How does one carry out an OSSE-like analysis for ML models? Replies to these questions suggested that very high-quality models need to be at the center of the test, but that the answers depend on the domain—some domains are mature enough for OSSEs while others might not be.

DATA FUSION AND ASSIMILATION

The Data Fusion and Assimilation Panel, moderated by Charles Norton, had six panelists: Mark Cheung of Lockheed Martin; Bernie Jackson of the University of California, San Diego; Slava Merkin of the Applied Physics Laboratory (APL) of Johns Hopkins University; Alex Chartier, also of APL; Tomoko Matsuo of the University of Colorado Boulder; and Eric Blasch of the U.S. Air Force Office of Scientific Research. The panelists were asked to address three questions:

- What are the new data assimilation/fusion approaches that will likely lead to improved space weather forecasting performance?
- Do you anticipate adequate data resources for these schemes? If not, how can data buys or other investments alleviate shortcomings?
- How can we quantify uncertainty in data assimilation schemes that use multi-source observations?

A discussion followed the panelists' presentations.

In the first presentation, Cheung addressed each of the three questions in turn. Concerning promising new approaches, he said that the large dimensionality of space weather problems requires efficient, scalable DA techniques, such as density estimation and data generation (e.g., normalizing flows), physics-informed neural networks, and neural network–based surrogate models. He said that the data resources are not adequate—either now or in the future—for modeling solar magnetic activity and its space weather impacts in the heliosphere; an adequate result would require multi-point measurements, potentially including solar polar magnetic field measurements. He added that the Solar Dynamics Observatory, the key data source today, does not have redundancy or continuity, and he said that it is important to now invest in future infrastructure, such as ngGONG or space-borne vector magnetographs. Data buys could help solve the problem if they would lead to lower costs of the magnetograph data.

On quantifying uncertainty, Cheung pointed to the power of approximate Bayesian computation methods and said that the model quality (or appropriateness) could be assessed by developing appropriate cross-domain, cross-source cost functions.

Jackson spoke about the short-term opportunity to improve global heliospheric analyses by taking full advantage of all (potentially) available data, including the Worldwide Interplanetary Scintillation Stations (WIPSS). A longer-term solution should combine ground-based remote-sensing data with space-borne heliospheric imagers.

Merkin talked about combining first-principle and data-derived approaches in global geospace modeling. Merkin said that the types of solutions will depend on the data types and availability and the physics-based models under consideration. Within the geospace, not only are the data sparse in spatial sampling but they are also unevenly sampled, with some variables measured in situ, others remotely sensed, and some not at all. He said that uncertainties arise both from missing physics and from challenges of modeling the vast range of spatial and temporal scales. Merkin expressed caution about uncertainty estimation through single-point data-model comparisons, as the uncertainty should be assessed using accuracy over a time window and metrics that weight physical relevance.

Alex Chartier asserted that the largest source of uncertainty comes from the electromagnetic and kinetic energy input into the ionosphere and upper atmosphere, which are crucial drivers of plasma convection and other (high latitude) space weather phenomena. Because these inputs have a high degree of temporal and spatial variability, they cannot be observed from a single vantage point; the best way to observe those parameters would be a low-Earth-orbit satellite constellation. While the 66-satellite Iridium constellation provides Active Magnetosphere and Planetary Electrodynamics Response Experiment (AMPERE) magnetometer data, other types of data needed include electric fields, particle precipitation, and plasma conductivity, ideally at high temporal and spatial resolution and well curated to remove biases and characterize uncertainties.

Matsuo advocated using systems approaches to unify observational and modeling capabilities, building on existing infrastructure (e.g., from the National Weather Service DA and ensemble weather forecasting facilities). While OSSEs can be useful for testing such systems, physics missing from the system description can give rise to degeneracies that make it difficult to determine whether the errors arise from the data or from the missing physics.

It is important to quantify not just model errors, Matsuo said, but also the "representativeness errors" that indicate systemic problems in data interpretation and handling (such as retrieval process uncertainties related to how sensor data are transformed into geophysical data products). Furthermore, it was pointed out that the space science community faces some self-imposed conflicts between the interoperability requirement of DA and the stand-alone science justification requirement of NASA missions.

In the final presentation, Blasch identified nonlinear, non-Gaussian evidential reasoning as a new technique that can be used to improve space weather forecasting. He suggested that the best way would be to use evidential reasoning as the last step in a process that otherwise relies on ML, which would speed up the system and also capture the uncertainty. Finally, data fusion with nonlinear, non-Gaussian DA (e.g., the unscented Kalman filter) can be used to make estimates of the states of nonlinear systems.

Blasch commented that the challenge in quantifying uncertainty is that there are so many different types of it—an ontology of uncertainty in the data fusion community identifies more than 50 sources of uncertainty. It is important to identify the types of uncertainty, be they absolute or relative, forward or inverse, epistemic or aleatoric.

Discussion

In the discussion, Norton asked about data fusion products that are not currently available and that are not directly observable. Generally speaking, even with its limitations, data fusion opens possibilities—especially multi-sensor approaches can help resolve new physical parameters. Successful examples include deducing the heliospheric magnetic field using ground-based Faraday rotation measurements of galactic sources combined with models of ionospheric total electron content, or constructing an extreme ultraviolet irradiance emulator from the Solar Dynamics Observatory (SDO) Atmospheric Imaging Assembly instrument using training data (fusion) from the SDO's Extreme Ultraviolet Variability Experiment.

Next Norton passed along the question of which are the most effective data with which to improve DA-based models. Matsuo noted that the ionospheric plasma eddy diffusion and conductance could both be useful additions, but anything that can constrain the models is helpful.

Chartier commented that data buys for space weather purposes are only possible if someone builds the instruments. To that end, he suggested to "think big" and communicate the need for specific types of sensors and instruments on constellations to the space companies, suggesting agreements that are mutually beneficial.

An extensive Zoom chat took place during the session. The discussion initially centered on solar data, with (unanswered) questions regarding data buys for solar magnetograms, and whether L4 and L5 views would be adequate (which is not yet clear). Ongoing efforts include combining modeling, DA, and available data to improve boundary inputs to global models, and DA with multi-view solar wind data, which relies on data that are at this point not operationally available. ML (with past measurements as input) could augment sparse data that could then be assimilated; however, it is not clear how independent the resulting data would be for statistical purposes.

For measurements of particles and fields, networks of small or nanosatellites might be capable of covering the spatial and temporal scales without the cost becoming prohibitive. However, as the correlation lengths and time-scales of interest are still open questions, the total number of necessary spacecraft remains unknown.

In any case, the challenge remains of enticing companies to build commercial off-the-shelf (COTS) science-grade instruments. OSSEs were once again mentioned as means to constrain the options.

In contrast with a statement by an earlier panelist, one chat participant expressed support for using JEDI as a framework, stating that it is interoperable with different forecast models. Another participant said that using the Faraday-rotation analysis of fields and electron density in the heliosphere and the ionosphere might suffer from noise created by near-Earth auroral kilometric radiation.

The Zoom chat discussions also covered the recent SpaceX incident in which more than 40 satellites were lost due to orbit maneuvers during a G1-level geomagnetic storm. Later analysis showed that while the storm itself was not remarkable, these conditions led to significant increases in the atmospheric neutral density. This complex event revealed problems in the operational Mass Spectrometer and Incoherent Scatter radar atmosphere model, highlighting the potential for problems during larger G5-level storms, given the challenges presented by this G1 storm. Given the data availability, the incident was seen as a good testbed for ML approaches.

ENSEMBLE MODELING

The Ensemble Modeling Panel, moderated by Mary Hudson, had six panelists: Sean Elvidge of the University of Birmingham, Eric Adamson of the Space Weather Prediction Center (SWPC) of NOAA, Dan Welling of the University of Texas at Arlington, Nick Pedatella of the University Corporation for Atmospheric Research, Kent Tobiska of Space Environment Technologies, and Edmund Henley of the UK Meteorological Office. The following questions were posed to this panel:

- The terrestrial weather community does multiple-model ensemble modeling. Is that practical for space weather in the near term?
- What data sources are needed but unavailable (proprietary, classified, etc.) that are hampering next steps? Do we work to get them available or can other methods (ML, data curation, fusion, etc.) suffice?

A discussion period followed the presentations.

Elvidge's general overview noted that ensembles can be used in a variety of ways: for DA, with multi-model ensembles (MMEs), or in uncertainty quantification. While the use of MMEs is in its infancy, it has demonstrated an ability to reduce errors, which has led to its use especially to reduce model-propagation uncertainty in DA. A key question in using MMEs is how to generate independent (orthogonal) ensemble members that are needed to estimate covariance matrices, to reduce propagation errors, and to ascertain that errors in individual contributions will effectively cancel. In closing, Elvidge noted that ensemble modeling does require an investment in computational resources and data storage, as the amount of data increases linearly with the number of ensemble members and with the number of different models.

Adamson described the state of ensemble model deployment at SWPC. At present, he said, while no ensembles are in operation, their potential to constrain uncertainties is recognized, and development work is ongoing in the solar/solar wind domain. The uncertainties in, for example, CME models are mainly observational and can be addressed by improved coronagraphs and new measurement vantage points.

Adamson returned to the two types of ensemble models, those comprising multiple realizations of the same model versus MMEs. As an example of the former, he showed a graph with predictions made by the Air Force Data Assimilative Photospheric Flux Transport (ADAPT) model, which is a 12-member ensemble with initial condition variation in supergranulation (Figure 5-3). Different initial conditions led to very different predictions for the arrival times of the CME.

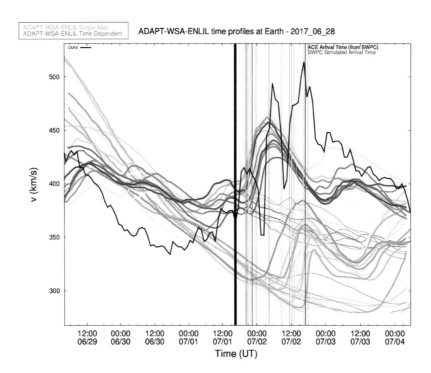

FIGURE 5-3 Variation in the Air Force Data Assimilative Photospheric Flux Transport (ADAPT) results due to differing boundary conditions for the solar wind speed at Earth orbit. Blue and green lines are different model runs, while the black line is the actual observation.
SOURCE: Eric Adamson, NOAA Space Weather Prediction Center, presentation to workshop, April 14, 2022.

SWPC has done little with MMEs, Adamson said, in large part because so much of the uncertainty is due to a lack of data. Although MMEs can be resource-intensive, they can be of interest in some cases—for example, in understanding how to effectively prune the ensembles, especially in the context of coupled-model systems.

Welling spoke about the use of ensembles in the geospace regime, particularly global magnetohydro-dynamics coupled to ionosphere and ring current models. At present, he said, the use of ensemble models is very limited in this area and is mostly of the proof-of-concept type. As the validation efforts are not well refined, the performance of ensemble-based approaches remains unclear. However, he continued, early work has shown that ensembles can increase the predictive skill of geospace models. In particular, studying model sensitivity to boundary conditions and missing or new physics, exploring the physics hidden in input parameters, and quantifying uncertainties across all inputs were mentioned as "low-hanging fruit."

To maximize the effectiveness of ensemble modeling requires a number of new resources, he said. Covering the driving by upstream L1 and near-Earth monitors and monitoring the impacts by a dense network of ground magnetometers, auroral imagers would make it possible to validate what Welling described as "one of our weakest points of modeling—auroral conductance and dynamics." Finally, he commented on the challenges in geospace DA: the data sampling needs to cover large volumes, and the processes described cover a range from very local to global. This often leads to point-source sampling that perturbs DA locally with undesirable global effects for the models. Assimilation of ionospheric electrodynamics, auroral observations, and ground magnetometers may be possible, although the viability needs to be assessed.

Pedatella spoke about the need to develop and evaluate new approaches for generating ensembles to address issues such as the ensemble not reflecting the uncertainty in the input parameters or in the model itself, or sensitivity of the model to initial and boundary conditions caused by chaotic dynamics. MMEs are widely used in weather and climate applications, and Pedatella saw opportunities for their use also in space weather. However, implementing MMEs for ionosphere–thermosphere applications will require new research as well as investments in the computing infrastructure, running the simulations, and distributing the results.

Tobiska described ensemble modeling of the thermosphere at the 18th Space Control Squadron of the U.S. Space Force. The High-Accuracy Satellite Drag Model (HASDM) is run near-continuously to predict thermospheric density for the coming 72 hours and used to inform satellite operations. There is a recognized need to reduce the absolute error in the model, Tobiska said.

Henley offered an operational perspective on the quantification of uncertainties in ensemble modeling. He began by noting that the operational needs are uncertainties due to current conditions—weather-of-the-day uncertainties—rather than uncertainty in performance against climatological references. As the classic ensemble approach to uncertainty is expensive, he recommended cheaper options, such as models that use simpler physics, reduced-order models with simplified internal dynamics, surrogate/emulator approaches as used in climate science, or ML-based models of the uncertainties themselves. These cheaper options may suffice for operational forecast needs.

Henley also made the point that it is important to communicate with the users to understand their needs and educate them on new opportunities. Resources spent on quantifying uncertainties are wasted if they do not inform the users' decision making, while providing that information in more helpful and targeted formats might increase their use. Terrestrial weather and climate forecasting rely on social science input to help identify users' needs, he said, including how to interpret uncertainties. As an example he mentioned Decider, a multi-criterion group decision-making support system that can help forecasters supply uncertainties that will be informative to a particular set of users.

Discussion

Hudson began the discussion period asking why ensembles of models should be preferred over the single model in the ensemble that is demonstrably the best. Pedatella responded that because it is rare that one model will always be the best, combining multiple models can ensure better results over a range of different situations. Elvidge added that not only is it often difficult to tell which model is best, but even if one is consistently better than the others, an ensemble of models will outperform the single best one.

To Hudson's request for a definition of ensemble models, Henley said it can be anything from changing something about the model (e.g., initial or boundary conditions) to altering the models being used. Tobiska said that the MME approach of using a spread of different models makes it possible to quantify the "uncertainty in our collective knowledge of the system." In the case of individual models providing probabilistic forecasts, Elvidge said that the ensemble output represents a probability distribution from which the collective uncertainty in the understanding of the system can be estimated.

In the case of applying DA or ensemble modeling to the magnetosphere, Welling said that distributed multipoint measurements can provide localized nudging without an overall large impact, but the success of the process will depend on the details of the assimilated information.

In the Zoom chat participants brought up the issue of MMEs versus ensemble DA versus a broader definition of a "perturbed physics, empirical, statistical, and machine-learning model." The point was made that MMEs are preferred over a single "best-performing" model, both because the different models, despite their performance, can bring useful information and because a combination of multiple models can outperform even the best single model.

From the operational perspective, single models rarely perform well under all conditions and scenarios. MMEs are useful in operational settings because in these settings it is difficult to know a priori which regime is present. The MMEs provide information about variance across models but not always information on their causes. One example is a solar wind ensemble model where the differences in individual models of the solar wind are small compared with those arising from a CME, indicating that MME for the solar wind component does not bring appreciable gains.

The Zoom chat touched on the lack of research of assimilating data (e.g., from ground magnetometers to magnetospheric MHD models); several comments cited the inherent difficulties in imposing observational constraints from the ionosphere to the magnetosphere as well as the fact that there are easier and perhaps more fruitful research topics available.

The chat participants noted that winds and temperatures from lower-atmosphere weather models are used for the forcing from the lower thermosphere and that MMEs of these weather models can capture the uncertainty to improve thermosphere-model ensemble generation. For whole-atmosphere DA, the driving from the lower atmosphere is incorporated from assimilating observations in the lower atmosphere, with some constraint in the mesosphere and lower thermosphere (MLT) provided by research-satellite data (i.e., TIMED/SABER and Aura/MLS). Wind estimates from meteor radars have been tested but not yet shown to be effective.

DATA AND MODEL RESOURCES AND CURATION

Another panel was devoted to data and model resources and their curation, with a specific focus on R2O2R (research-to-operations and operations-to-research) efforts. Moderated by committee member Anthea Coster, the panel had six presenters: Carrie Black of NSF, William Schreiner of the University Corporation for Atmospheric Research, Jack Ireland of NASA, Rob Redmon of NOAA, Masha Kuznetsova of NASA's Community Coordinated Modeling Center, and Alec Engell of NextGen Federal Systems. The following questions were asked of this panel:

- Is a more coordinated/sustained data curation effort needed to support R2O2R?
- What data sources are needed but unavailable (proprietary, classified, etc.) that are hampering next steps? Do we work to get them available or can machine learning, data curation, etc., take care of it, and how?

Black offered an overview of NSF data programs in space science, which are served by two NSF divisions, the Division of Astronomical Sciences and the Division of Atmospheric and Geospace Sciences noting that the agency is on the research end of research-to-operations. The unmet data infrastructure needs of that community include an easy-access, user-friendly data infrastructure, file and data standardization, documentation of data for record keeping and end users, and the development of one or more data repositories. NSF supports solar and space physics data systems, such as the Madrigal Database for archival and real-time data from upper atmospheric science instruments and the Community Coordinated Modeling Center for space research models. Although NSF does provide funding for the CubeSat program, Black said that most of its support is focused on ground-based data acquisition.

While NSF has the FAIR (findable, accessible, interoperable, reusable) data policy in place, Black said, data management plan requirements are not uniform, and the reporting of these topics lacks guidance or enforcement. Noting that the decadal survey strategy considers the "data and computing infrastructure needed to support the research strategy and the long-term utility, usability, and accessibility of acquired data," she suggested that community participation in the decadal survey might be augmented by developing curricula for data resource use and open access and by forming working groups to address some of the topics.

Schreiner spoke about data curation and analytics issues specifically related to radio occultation data. He offered a historical overview of missions providing radio occultation data, beginning with GPS/MET (for GPS Meteorological experiment) in 1995 and culminating with COSMIC-2 (Constellation Observing System for Meteorology, Ionosphere, and Climate 2) and today's commercial data buys from global navigation satellite systems. Commercial radio occultation data are an integral part of assessing the state of the ionosphere both now and in the foreseeable future; COSMIC-2 combined with the commercial buys can provide global coverage with some gaps poleward of ±40 degrees latitude. However, the radio occultation data needs standard data and metadata formats, and the large data volumes and complex models require cloud-computing environments. In general, more coordinated and sustained data curation efforts as well as community-developed space weather assimilation models are needed to support R2O2R. Concerning data needs, Schreiner pointed to low-latency access, gridded products, standardization of data formats, and data-proximate cloud computing environments that would enable science applications and assimilative space weather models.

Ireland spoke about NASA's Heliophysics Digital Resource Library (HDRL), which makes heliophysics data, tools, and services available to the broader community, with a focus on the research side of R2O. The high-level strategy behind HDRL can be broken down into four main categories of goals: preserve (i.e., maintain and improve existing archives), discover (support researchers' efforts), explore further (enable big data research by bringing together high-end computing and large data sets), and share and publish (support collaborative research and publishing platforms). Recognizing the increasing volume and diversity of solar data, HDRL wishes to help users accelerate heliophysics research. To do so, NASA has set out four broad strategies: enabling open science, lowering current barriers to doing research, implementing new critical capabilities, and being responsive to changing community needs. The science infrastructure for these strategies has four components: the Heliophysics Data and Model Consortium, the Space Physics Data Facility, the Solar Data Analysis Center, and various collaborators, such as the Community Coordinated Modeling Center (CCMC) and the Center for HelioAnalytics. In other words, the path forward will involve both NASA internal deliberations and community inclusion and direction.

Redmon suggested that one way to support R2O2R would be to use new programs such as NASA's HDRL (described earlier by Ireland) and NOAA's Space Weather Follow On (SWFO) program to take advantage of interoperable data and metadata standards (e.g., SPASE, the Space Physics Archive Search and Extract) and existing interfaces (e.g., the Heliophysics Application Programming Interface and the Deep Space Climate Observatory [DSCOVR]) to develop benchmark data sets similar to augment climate data records. Another strategy would be to take advantage of collaborations among government, academia, and industry, such as Earth Science Information Partners.

Redmond called for continuing efforts to bring existing but unavailable data to the research community to drive future R2O2R, for example, via the Space Weather Operations, Research, and Mitigation (SWORM) Subcommittee. Notably, he said, NOAA is working to migrate all its data holdings to the cloud and that it is embracing the FAIR policy. Further ideas included tackling modeling challenges with crowd-sourcing (with prize money) and applying ML or AI to anonymize proprietary or protected data for R2O2R use.

Kuznetsova described the Community Coordinated Modeling Center and its capabilities. She focused on the role that the "shared proving grounds" between CCMC and the SWPC play in R2O2R (Figure 5-4), providing, for example, researchers access to operational data streams, model inputs, and simulation outputs through the CCMC.

Kuznetsova listed a number of steps that CCMC is planning to take to improve its processes and products, such as establishing and following best practices for on-boarding and implementation, improving the quality of simulation archives, improving the robustness and speed of simulations, and a move toward "plug and play" models as part of open-source pilot projects. The CCMC also wishes to increase involvement of the modelers in transitioning models to CCMC and implementing GPU-ready code, but the funding for these items was not addressed.

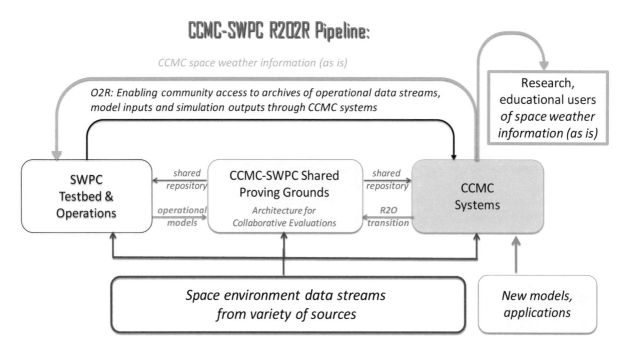

FIGURE 5-4 A research-to-operations/operations-to-research (R2O2R) pipeline between the operational Space Weather Prediction Center (SWPC) and the Community Coordinated Modeling Center (CCMC).
SOURCE: Masha Kuznetsova, NASA Community Coordinated Modeling Center, presentation to workshop, April 13, 2022.

Engell promoted coordinated high-level infrastructure and execution as a way to support R2O2R, and he called for the definition of best engineering practices drafted by a focused team with outside expertise. He listed a number of specific challenges, including the diversity of data, lack of curation standards, limited availability of operational and research models, and the scarcity of data from both space- and ground-based instruments. He pointed out the many data curation and R2O2R efforts that are progressing in parallel (e.g., NASA's HPDE/HPCLoud, the SWPC Testbed, the Space Force's Space Domain Awareness Environmental Toolkit for Defense [SET4D], the Air Force's Unified Data Library, NOAA AI efforts), and he characterized this as a positive development. The Earth sciences community is developing substantial infrastructure, specifically Pangeo,[2] as a community platform for big data geoscience, and the space weather community could learn from that effort and work toward building a "Panhelio" facility.

Discussion

The Zoom chat included significant discussion centered on concerns about guaranteeing the longevity of data curation and archiving. The potential role of the World Data Center was mentioned, and its limited role in space weather data curation and archiving. Cloud resources was seen as preferable to facility-based storage systems, but some expressed concerns about the longevity of that solution.

---

[2] See Pangeo, "Hompage," https://pangeo.io, accessed August 10, 2022.

The very definition of "long term" was discussed in light of 3- to 5-year planning and proposal funding cycles, NSF facility hardware divesting requirements (which do not include long-term data archival), and the challenges of archiving historical data, including data owned by principal investigators.

With regard to data buys, concerns were expressed about quality control, longevity, reliability, and user access. NOAA's current radio occultation data buys are made available to the research community with a 24-hour delay, but the committee heard nothing about standards or practices regarding these data.

One commenter said that radio frequency data from the solar corona are key to space weather research and forecast operations, but their archiving, curation, and continued acquisition were not addressed during the workshop, and the NSF support focused on these observations is small.

Similarly, there was concern about the data gap that will be left when the Defense Meteorological Satellite Program (DMSP) and the Polar Operational Environmental Satellite (POES) programs are retired. This will especially affect the ability to monitor spacecraft charging in the auroral zone, forcing operators to resort to models for anomaly resolution.

Further discussion regarding the archiving of SWPC operational models and forecasting products revolved around NOAA's National Centers for Environmental Information (NCEI), which is intended to archive all SWPC operational forecasts and models. It was unclear what exactly NCEI is mandated to archive, and that lack of clarity mirrors the current lack of resources, especially when coordinating across agencies.

The topic of interagency coordination becomes even more problematic when considering data sharing agreements with international missions. For NASA, such agreements would involve an office other than the science mission directorate responsible for the mission to direct the HDRL to act according to the agreement. The international aspect also points to the question of whether better and more sustained data curation would support the use of non-NASA data sets in NASA proposals; this question was viewed by the panelists as a high-level policy issue.

6

Research Infrastructure

Key Themes

In the sessions devoted to research infrastructure, the following key themes emerged from the presentations and discussions:

- It is important to develop architectures that will bridge the major gap that exists between solar and heliospheric research and geospace research, fusing research in both areas.
- Three-dimensional coverage of the Sun and heliosphere, with measurements that have high time and energy resolution and large dynamic range, are needed to close the observational gap of the 3D structure of transient phenomena. One solution would be to introduce out-of-the-ecliptic architectures that would provide three-dimensional reconstruction of transients traveling toward Earth, including their magnetic field structure.
- More multi-point distributed measurements throughout the system are critically needed for science, in particular for improving empirical and physics-based models.
- Key information about the Sun and the heliosphere is needed in near real time. Various pieces of existing infrastructure could be repurposed to serve research needs, instead of building new infrastructure. An example of such dual use infrastructure would be commercial and Department of Defense communication networks to provide data downlink and reduced-latency services.
- A systems-level approach focused on multi-scale measurements is key to addressing gaps and sparse coverage of the observations, and to addressing the multi-scale nature of various space weather phenomena.
- It may be possible to design both individual instruments and spacecraft so that they can be used effectively for both research and space weather observations. To do so, it is crucial that the instrument design maintains multiple analysis and operation pathways as viable, so that some can be devoted to research while others focus on collecting space weather data.
- New capabilities are needed for managing the larger data volumes now being produced, including techniques for processing data onboard the spacecraft.

As capabilities have steadily increased in areas such as spacecraft development, launch systems, modeling and data analytics, instrumentation, and commercial data product services, the space weather community's view of the future of science mission infrastructure and operations has changed. Emerging trends across these areas will play a major role in re-shaping the architecture that supports efforts to better understand the Sun–Earth system and the impacts of space weather in space and on ground. Furthermore, the ways that information flows from research to operations and operations to research (R2O2R) are also evolving.

Two panels specifically addressed the issues of the architecture needed to support space weather research and operations. One panel was focused on the Sun and heliosphere, while the other addressed the magnetosphere and ionosphere–thermosphere–mesosphere system. In particular, these presentations addressed the part of the statement of task that asked the committee to

- *Take into account the results of studies, including NASA's space weather science gap analysis (part of the NASA Heliophysics Division's Space Weather Science Application program) and the NSF Investments in Critical Capabilities for Geospace Science (2016), to identify the key elements needed to establish a robust research infrastructure*

The panelists in each of the two panels were asked to address three key questions:

- What are the novel observational/model architectures/technologies that are not yet being used?
- What is needed to build a fluently (i.e., smooth and effective) operating architecture from the multi-source/multi-organization observational base that currently exists?
- What are the government/private resources (data and platforms) that could be used but that are not being used right now?

This chapter summarizes the discussions on those topics and is divided into two main sections, the first on the Sun and heliosphere, and the second on the magnetosphere and ionosphere–thermosphere–mesosphere.

THE SUN AND THE HELIOSPHERE

The New Architectures: Solar and Heliosphere Panel was moderated by committee member Dan Baker. The panelists were Justin Kasper of BWX Technologies, Nicole Duncan of Ball Aerospace, Tom Berger of the University of Colorado, Sue Lepri of the University of Michigan, and Angelos Vourlidas of the Applied Physics Laboratory of Johns Hopkins University.

The panelists used the three key questions above as guidance to explore new architecture issues within their areas of discipline expertise. The topics they discussed included multi-spacecraft small satellite constellation mission architectures; innovative partnerships, contracting, and procurement mechanisms across government agencies; space and ground infrastructure capabilities; and enabling technologies. In addition, the panelists covered new observational vantage points, measurement concepts and their connections, observational gap analysis, and common instrument suites, all in support of advancing space weather–related solar and heliospheric science and research.

The following summarizes the key points from the panel presentations and discussions.

Programmatic, Policy, and Contracting Innovations for New Observations

In Kasper's presentation, Milestone-Based Reimbursable Space Weather Missions Leading to Department of Commerce or Department of Defense Data Purchase Agreements, he advocated for the development of a commercial data-buy architecture to produce heliophysics data products. Given the fact that launch services, data services, spacecraft systems, and operations are all becoming more commoditized, such an approach would facilitate advanced heliospheric space weather observations at substantially lower cost—and faster deployment time—than existing contractual mechanisms.

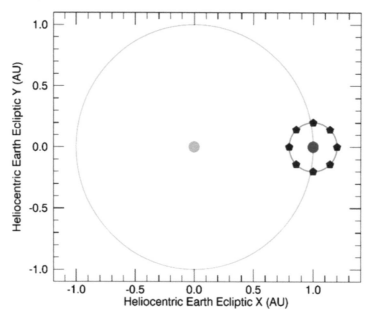

FIGURE 6-1 An advanced warning multi-spacecraft mission concept.
NOTES: The figure captures the appearance of the constellation encircling Earth at a distance of 0.2 AU from Earth.
The spacecraft closest to the Sun would give double the lead time as compared to current space weather monitors
at L1 at 0.1 AU Sunward of Earth.
SOURCE: Justin Kasper, BWX Technologies, presentation to workshop, April 13, 2022.

Kasper presented an example mission architecture that could benefit heliospheric science simultaneously with providing a milestone-based reimbursable space weather monitoring capability and data
purchase agreement structure for the Department of Commerce and Department of Defense (DoD)
(Figure 6-1). An eight-spacecraft constellation platform around Earth in a retrograde orbit at 0.2 AU distance from Earth could give advance warnings of transient arrivals with lead times that are longer than
can be accomplished with L1 alone. Carrying simple plasma, magnetometer, and radiation sensors, this
mission design would give multiple lines of sight and much greater confidence in understanding the
magnetic field and probabilities of affecting Earth. At a cost of around $100 million, this infrastructure
would allow airlines to take more effective action in response to transient arrivals by having a minimum
of 25 hours advance warning with three-dimensional information on shock fronts, flux rope structure,
and energetic particles.

Kasper argued that such an architecture could be provided commercially today. Companies could
make a profit by selling data through milestone-based reimbursable data purchase agreements with an
augmentation from the National Aeronautics and Space Administration (NASA) space weather budget. The
challenge in setting up the system is that an interplanetary mission could represent a large upfront cost for
a company, as it requires more resources (is more expensive and takes longer from design to operations)
than Earth-observation counterparts. A possible solution to this problem would be to develop a mechanism

for commercial data buys where NASA, the National Oceanic and Atmospheric Administration (NOAA), or other interested agencies could enter a milestone-based commercial data acquisition agreement covering the period from instrument development to launch and data production.

In her presentation Enabling Solar and Heliospheric Novel Architectures, Duncan argued that a joint long-term plan and sufficient budget would be needed to enable the sort of novel mission architectures that Kasper described. Duncan mentioned that one approach would be for the U.S. Congress to fund the Promoting Research and Observations of Space Weather to Improve the Forecasting of Tomorrow (PROSWIFT) Act, without affecting other areas of interest, to move the community forward.

Combining scientific and space weather goals has also raised concerns within the science community. From the mission formulation perspective, new technologies can help missions to perform accounting for both scientific and operational needs. Similarly, there are opportunities through the research-to-operations/operations-to-research (R2O2R) Framework for gaining understanding of the operational requirements on the mission design. However, the community concerns about prioritization of scientific versus space weather goals need to be addressed and may require greater flexibility in mission architecture planning.

In addition, mechanisms to transition science mission data to operational use, including the reduction of data latency, are key concerns. The varying success in operationalizing science data could be addressed by greater standardization, and user barriers could be removed by improving data access through standardization and graphical user interfaces. In considering architectures to operationalize mission systems, Duncan said, an industry-prime public–private partnership model has been successful (e.g., with defense customers). In this model, the agency defines the top-level goals, and industry competes to come up with mission definition architectures. As an example, Ball Aerospace uses a mission analysis suite called MOSAIC that optimizes the mapping of top-level goals to mission design (Figure 6-2).

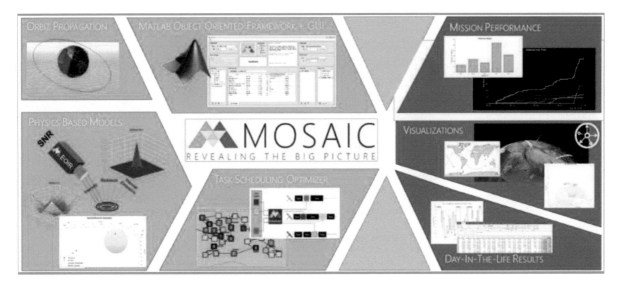

FIGURE 6-2 The Ball Aerospace MOSAIC mission analysis suite optimizes and streamlines mission design from top-level mission goals.
SOURCE: Nicole Duncan, Ball Aerospace, presentation to workshop, April 13, 2022.

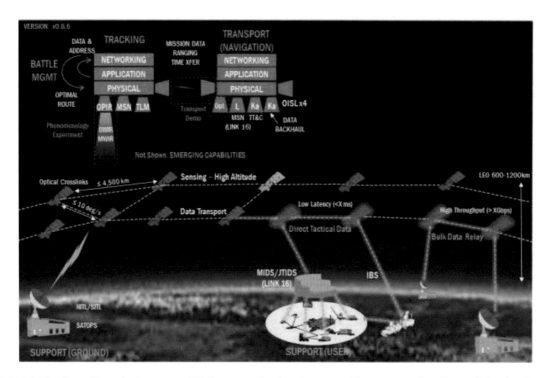

FIGURE 6-3 The Space Domain Awareness (SDA) communications layer provides an example of how a hybrid architecture of space-based and ground-based assets can provide real-time data.
SOURCE: Nicole Duncan, Ball Aerospace, presentation to workshop, April 13, 2022.

Following such commercial practices may lead to lower costs by enabling minimal oversight, reducing deliverables, and using industry-defined mission assurance practices, existing capabilities, and standard interfaces.

Communications infrastructure is another key architectural area where commercial and DoD communication networks could be used to provide data downlink and reduced latency services instead of building new infrastructure dedicated to space weather. For example, expanding NASA's Deep Space Network (DSN) and increasing launch vehicle rideshare opportunities in collaboration with the Planetary Science Division at NASA could be an effective means of broadening future space mission architectures for heliophysics (Figure 6-3).

Spacecraft constellations offer new opportunities for space weather research and operations, but they may require changing the processes of how missions are currently designed: Rather than building a single large constellation mission at once, one could deploy constellations over time, which would allow for periodic refreshment, increased lifetime, and rideshare opportunities. However, it will be necessary to think about how to structure contracts and procurements—using multiple contracts and minimizing the time between unit builds will be important to achieving sustainable, long-term science observations with resiliency and cost control (see Figures 6-8 and 6-9). In addition, the architecture needs to account for constellation-specific technologies, such as autonomous formation flying in the near-Earth environment, which can affect space traffic management. While it is unclear exactly what the next-generation mission architecture will look like, the policy and resources issues must be considered early.

Finally, those designing the constellation missions will need more guidance on assurance expectations, approaches to testing, space traffic management, and the need for on-board artificial intelligence and machine learning. The framework for technology readiness level (TRL) maturation and qualification is as yet not well understood for the constellations; these issues must be addressed to ensure successful heliophysics constellation architectures.

New Architectures to Overcome Observational and Modeling Deficiencies

In the next presentation, 4π HeliOS Mission Concept to Advance Space Weather Research and Operations, Berger discussed how current observations of the Sun are very limited (Figure 6-4), which inherently limits the accuracy of all existing space weather models. Currently, solar observations are confined largely to the Sun–Earth line. Even when a map of the full Sun is produced using the Sun's rotation, the map does not represent an instantaneous state of the Sun at any given time but instead is a convolution of temporal evolution and spatial information.

The view from the ecliptic plane contains limited information, as only about a third of the solar surface is visible at any given time. To estimate the synoptic solar flux, the measured values are extrapolated to the total solar output (see Figure 6-4), where the full modeling process introduces a forecasting delay

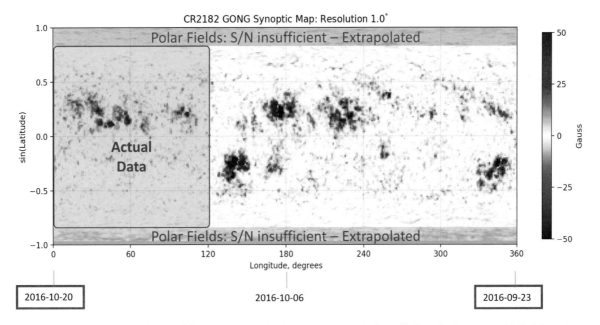

FIGURE 6-4 The current limited view of the Sun. From a single vantage point in the ecliptic only about one-third of the sphere is measurable at any one time, and the polar fields are not measured well. Solar wind modelers and forecasters routinely multiply synoptic map flux values by ad hoc factors of 2–4 to match measured conditions at L1.
SOURCE: Tom Berger, University of Colorado Boulder, presentation to workshop, April 13, 2022.

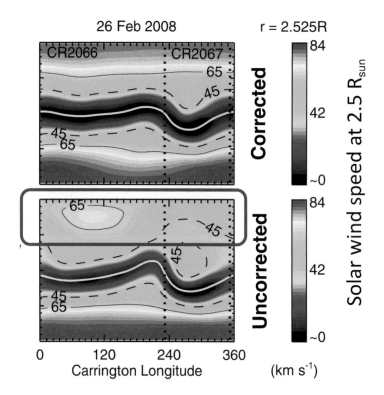

FIGURE 6-5 Polar extrapolations are very approximate where the North Pole fast wind is completely missed by uncorrected polar field extrapolation.
SOURCE: Tom Berger, University of Colorado Boulder, presentation to workshop, April 13, 2022.

of greater than 10 hours. Furthermore, inaccuracies in the polar extrapolations, which are used as model input, lead to very poor values for the solar wind speeds, which in turn compromises arrival time estimates for coronal mass ejections (CMEs). An example of the impact of the gap in observations at high solar latitudes is demonstrated in Figure 6-5, which shows that the fast wind from the Sun's northern polar regions is missed by the uncorrected polar field extrapolation.

The 4π HeliOS mission concept (Figure 6-6) would advance space weather research and operations by providing full solar coverage more than 80 percent of the time over the mission lifetime. The architecture consists of two solar–polar and two solar–quadrature (ecliptic plane) spacecraft, all of which are three-axis stabilized imaging platforms. The addition of an L1 mission would increase coverage along the Sun–Earth line over a 10-year mission lifetime. The solar–polar orbital inclination should be at least 60 degrees, with greater than 80 degrees desired. The solar–quadrature elements would be placed between ±90 and ±120 degrees, without requiring the use of the L4 or L5 points. All spacecraft would have perihelia of about 1 AU with a common instrument suite to facilitate intercalibration and redundancy.

The HeliOS instrument suite would combine in situ and remote sensing instruments. The remote sensing components would include a solar spectral irradiance monitor, Doppler vector magnetograph, an extreme ultraviolet (EUV) imager, and a white-light coronagraph. The in situ instruments would include

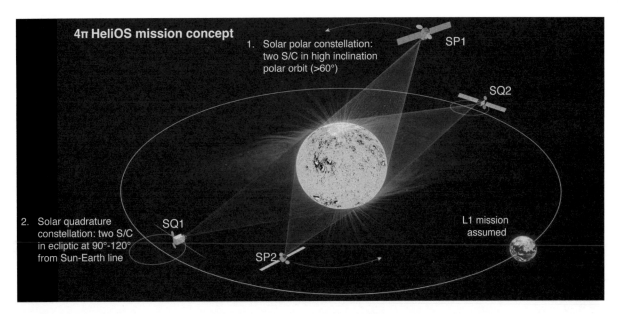

FIGURE 6-6 The 4π HeliOS mission concept architecture to address the most critical missing piece in space weather: full-Sun observations.
SOURCE: Tom Berger, University of Colorado Boulder, presentation to workshop, April 13, 2022.

a magnetometer, a solar wind plasma analyzer (Faraday cup), and an energetic particle suite. Additional capability would come if it were possible to include additional instruments. For example, the addition of Golf-NG (Global Oscillations at Low Frequency-New Generation) would allow detection of the solar gravity and g-modes, and thus probe the rotation rate of the Sun's core.

As envisioned, this mission concept still has significant technological challenges, including acquiring terabytes of data per day in near real time and using optical communications with associated deep space optical infrastructure to relay the data to Earth. Furthermore, if a single launch vehicle is to support both north and south polar spacecraft deployments, it must be equipped with ion or nuclear/thermal thrusters and be compact and lightweight enough to get out of Jupiter's gravity well. Autonomous spacecraft and data systems are a must to support helioseismology, which requires 100 msec-level accurate timing across the constellation. The mission complexity overall means that artificial intelligence and autonomy would be required.

In Lepri's presentation, Perspectives from the Heliosphere: ICMEs, SEPs, Suprathermal Particles and the Nature and Structure of the Solar Wind, she emphasized how observational gaps related to ICMEs (interplanetary coronal mass ejections), shocks, and the bulk solar wind limit the space weather forecasting/nowcasting capability and the understanding of how to derive key parameters, such as time of arrival, duration, and the geo-effectiveness of the structures at Earth. An example is show in Figure 6-7.

The key features to detect include the ICME structure, extent, and magnetic topology. Similarly, it is necessary to know the time of arrival, the boundaries, the duration, and the plasma and magnetic properties of solar wind structures, such as high-speed streams (HSSs), stream interaction regions (SIRs),

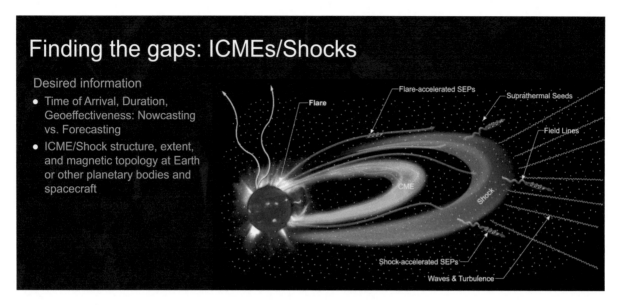

FIGURE 6-7 Finding the gaps: Interplanetary coronal mass ejections (ICMEs)/shocks.
NOTE: SEP = solar energetic particle.
SOURCES: Susan Lepri, University of Michigan, presentation to workshop, April 13, 2022; image from M.I. Desai and D. Burgess, 2008, "Particle Acceleration at Coronal Mass Ejection–Driven Interplanetary Shocks and the Earth's Bow Shock," *Journal of Geophysical Research* 113(A9), https://doi.org/10.1029/2008JA013219; copyright 2008 by the American Geophysical Union.

and corotating interaction regions (CIRs). Key open questions include the evolution of these structures as they travel through the heliosphere, including effects of preconditioning (previous events and prior state of the solar wind into which the structures expand) and the influence of the solar cycle phase. Addressing these questions will require monitoring the three-dimensional heliosphere with the ability to identify these phenomena.

Lepri said that there are significant gaps in existing in situ measurements as well as in the current ability to connect remote and in situ measurements. Enhanced observations and modeling efforts are needed to connect remote observations (line of sight integrated) to in situ (point) measurements. There are also limitations in the current instrumentation: most of the currently operative instruments cannot capture sufficiently high solar wind speeds or proton fluxes to accurately record the largest events (solar wind speeds may get up to 3,000 km/s). While heavy ion composition has shown promise in identifying different solar wind types, the lack of real-time monitoring of the ion composition adversely affects the quality of space weather nowcasting.

Heavy ion composition, which is a significant gap in current in situ plasma measurements, plays a key role in differentiating HSSs from ICMEs, in determining the internal structure of the ICME, and in constraining models of CME initiation, energization, heating, release, and propagation. These observations are essential to nowcasting and to quantifying the total mass and pressure of a space weather event. Composition information is now being incorporated into models, but validation data are still scarce. If composition measurements were available, they could be a key discriminating marker for machine learning (ML) algorithms used to identify and predict ICMEs, and can thus improve the ML methodology effectiveness.

Suprathermal particles are the leading indicator of shocks, and provide 5-72 hours of warning, depending on the strength of the shock. As suprathermal particles are also the seed population for solar energetic particles (SEPs), they provide a great forecasting metric. However, being less abundant than the bulk of the solar wind, their observation is more challenging. Many current particle instruments lack the dynamic range or sensitivity to cover the many orders of magnitude required to resolve the suprathermal component. The instruments are also not able to resolve smaller-scale composition and structural changes, because of the longer integration periods required.

Ultimately, the goals of observing the suprathermal and heavy ion components should be to enhance predictive capability, understand the background solar wind, and build a climatology of solar wind parameters. Mission architectures to support those goals would include distributed systems both in and out of the Sun–Earth line. Lepri listed a number of measurement priorities, including in situ plasma and energetic particle (composition) measurements at various distances ranging from 0.7 to 1 AU to track the flow and spatial variations occurring en route. Such a distributed network could also be achieved by using instruments hosted on planetary missions. Other priorities include complementary measurements across platforms to combine remote and in situ measurements, and getting long baselines and uniformity by sustained observations producing long-term data series.

Achieving this enhanced capability will require overcoming significant challenges. New capabilities (high-resolution proton and electron measurements, heavy ion composition, suprathermal particles, SEPs, and magnetic field) will be required to determine the impacts of smaller-scale structures. This, in turn, will require increased telemetry and real-time data downlinking capability, both of which may need novel onboard processing or advances in ground-based data-analysis tools.

Lepri used particle instruments as an example of challenges faced when increasing instrument requirements: higher resolution requires larger fields of view or geometric factors. Extending the mass, charge, or energy range of a measurement typically requires larger sensors, while the necessary instrument miniaturization is especially challenging for those needing high-voltage power supply. Each mission and architecture design must decide on the trade-off between the value of smaller, simpler instruments and the cost of more complex and capable sensors.

In the next presentation, Novel Observational/Model Architectures/Technologies to Fill Gaps in Observations of Solar Transients, Vourlidas followed the themes outlined in the NASA heliophysics gap analysis report by discussing the gaps in solar observations, starting with the observations of solar transients: CMEs, high-speed streams, shocks, and associated energetic particles. A key theme in his presentation was the lack of effective three-dimensional coverage of how the transient events evolve over time. Two key measurement challenges are the lack of coverage of events, and the scale mismatch of the observing techniques, with the in situ techniques having higher temporal resolution but sampling at just a single point, while imaging provides broader coverage but with spatial resolutions that are 1,000 to 10,000 times coarser.

Vourlidas emphasized the importance of taking a systems-level approach focused on multi-scale measurements as a way to address gaps due to sparse coverage and the multi-scale nature of the phenomena of interest. Such a system was discussed in the Berger presentation (Figure 6-8). A strategic plan and systematic approach are needed to build up such a system in response to the gap analysis report. The key new elements of this architecture are (1) that the platforms would be distributed (rather than monolithic), (2) the platforms at the Lagrange points would be in "quadrature" (four spacecraft rather than the currently used single vehicles), and (3) the spacecraft would carry only focused payloads (smaller numbers of sensors) (Figure 6-9). Novel approaches, such as the use of CubeSats and in situ constellations, could address some of the sampling issues. Communications will present a challenge, particularly for the envisioned far-side sensors. (See also the Berger presentation above.)

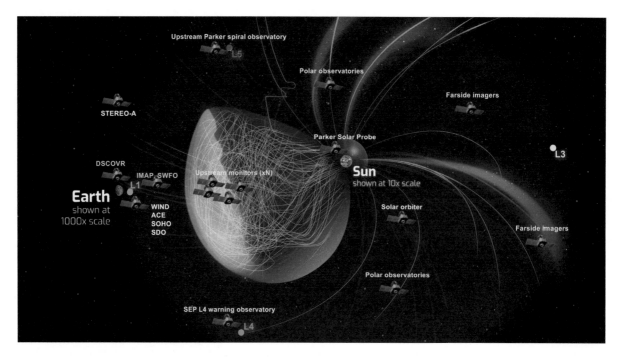

FIGURE 6-8 System-level architecture to fill observational gaps.
NOTE: Acronyms defined in Appendix D.
SOURCE: Angelos Vourlidas, Johns Hopkins University Applied Physics Laboratory, presentation to workshop, April 13, 2022.

Vourlidas spoke about the importance of developing architectures that bridge the major gap that exists between solar, heliophysics, and geospace research. In bridging the gap from the solar wind to geospace, care must be taken to implement a "peri-geospace" architecture that would include a system of L1-Earth "cyclers." These dedicated platforms would provide sampling of solar wind conditions off the Sun–Earth line and address the transient scale problem important for space weather.

Finally, Vourlidas commented that user requirements as of today are not sufficiently well defined to determine the measurement architecture. Developing a cost-effective strategy will require research and a modeling of the system (through observing system simulation experiments [OSSEs]) to determine the most effective measurement strategies. It is likely that new partnerships are needed to achieve these goals.

Interactive Panel Discussion

After the panelists finished their presentations, they took part in a broad-ranging discussion with questions from the viewing audience. Three major needs emerged from the discussions:

- More complete coverage of the Sun (i.e., from all directions).
- More comprehensive measurements, including high-resolution (in both time and energy) measurements and measurements with large dynamic range.
- New capabilities for managing the data volumes being produced, including techniques for processing data onboard the spacecraft, and autonomy for operations and data collection. A recurrent theme was the need for some key information to be provided in near real time.

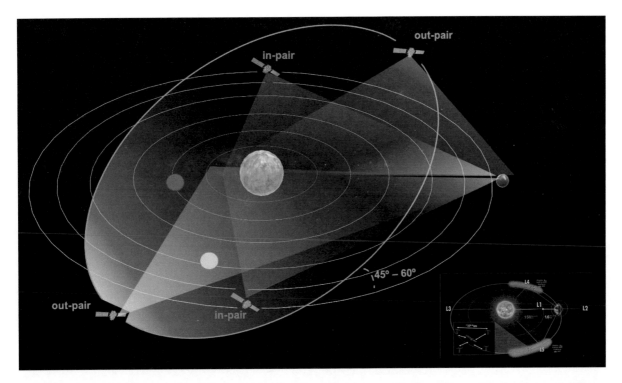

FIGURE 6-9 Out-of-ecliptic architectures for true three-dimensional reconstruction of incoming transients enabling a quantum leap in space weather forecasting capability.
SOURCE: Angelos Vourlidas, Johns Hopkins University Applied Physics Laboratory, presentation to workshop, April 13, 2022.

While the panel did not explicitly address costs, the requirements they outlined suggested that new ways of managing programs will be needed to increase the efficiency of the building and operations of constellations as well as the efficiency of data delivery. Constellations built by multiple partners and in phases could support development of sustainable mission architectures constructed in affordable steps. Enabling data buys was viewed as a viable alternative to the traditional NASA/NOAA-funded data collection.

The discussion also touched on public–private partnerships for data provision. Two main points were made: First, industry does not like to accept risk for data production when the customer is basically one organization (NOAA or NASA, depending on whether the data are for monitoring or for science). Second, NASA and NOAA follow different procurement and implementation practices than those used by industry. Industry practices are purported to yield lower costs at low risk by following standard, proven practices and eschewing unnecessary documentation and reviews. Sharing risk would require changes in the current NASA and NOAA procurement process. Finally, the mission class system followed by NASA does limit flexibility, especially when measurement continuity is desirable. A possible solution would be to address the funding profile constraints with fractionated or heterogeneous approaches that would allow NASA to match science and operational data collection needs by adjusting the makeup of the elements of the heterogeneous constellation or providing a needed measurement on a rideshare or hosted payload. It was also suggested that there may be ways to transition a science mission to an operational one once the prime science phase is completed, even if there are significant differences in the measurement requirements; however, these issues were not explored further.

Panelists also raised some concerns about public–private partnerships. In particular, broad industry involvement requires a clear understanding of the return on investment (ROI) and careful justification to the shareholders—examples were quoted where commercial data production has ended due to lack of ROI. It was also questioned how the R2O2R cycle would operate in such a setting, or how the funding profiles would be established and managed.

The discussion on mission architectures raised a question on how mission planning changes when considering a design that moves the research-to-operations (R2O) process up front, and how the OSSEs may be used to explore such questions. Furthermore, the OSSEs can be used to study how the constellation sizes affect science closure, the value of measurements for operational nowcasts and forecasts, and the orbits where measurements are most needed. The panel generally agreed, however, that new architectures do not need hundreds of spacecraft in deep space, just dozens in critical places.

Finally, the panel discussion continued the previous discussion on the need for new types of measurements and novel instruments, with general agreement that developing new instruments and technologies is best done in a university setting, which offers opportunities for innovation and workforce development through student involvement.

MAGNETOSPHERE, IONOSPHERE, AND THERMOSPHERE

The New Architectures: Magnetosphere–Ionosphere–Thermosphere Panel was moderated by committee member Dan Baker. Its panelists were Phil Erickson of the Massachusetts Institute of Technology's Haystack Observatory, Brian Anderson of the Applied Physics Laboratory of Johns Hopkins University, Robyn Millan of Dartmouth College, Katelynn Greer of the University of Colorado, and Erik Babcock of SpaceX. Asked to consider the same three key questions as the New Architectures: Solar and Heliophysics Panel, the members of this panel followed the first one by discussing how capability enhancements influence ways to develop new missions. The specific topics discussed by this panel included the role of ground-based research-grade instruments and nontraditional data sources, commercial data acquisition, data fusion and heterogeneous ground-space-based multi-point distributed measurements, and the impact of space weather and situational awareness on the ability to deploy spacecraft safely in low Earth orbit.

Infrastructure and Data Systems for New Observations

Erickson's presentation was titled Ground-Based Instruments: Considerations for Usable Space Weather Data. He spoke about the role of ground-based research-grade instrument infrastructure and the considerations and challenges involved with supplying usable space weather data. For example, incoherent scatter radars (ISRs), such as the European Incoherent Scatter Radar (EISCAT 3D), are designed with flexibility in mind and have many modes, which makes it challenging to incorporate the data into a standard space weather feed.

Nevertheless, he considered that it is possible to design instruments to serve both research and space weather needs. Ideally, research instruments would be designed from the start to also produce space weather feeds, and all instrument designs would keep multiple analysis and operation pathways viable to allow both flexibility for research as well as uniformity and reliability for operations. For example, space missions addressing the science of space weather deliver a series of data levels, from basic (Level 1) to various derived data products (Levels 2, 3, 4), of which Levels 3 and 4 could be usable for operations. He concluded that operational space weather products need well-characterized measurement fidelity and uncertainties as well as low latency.

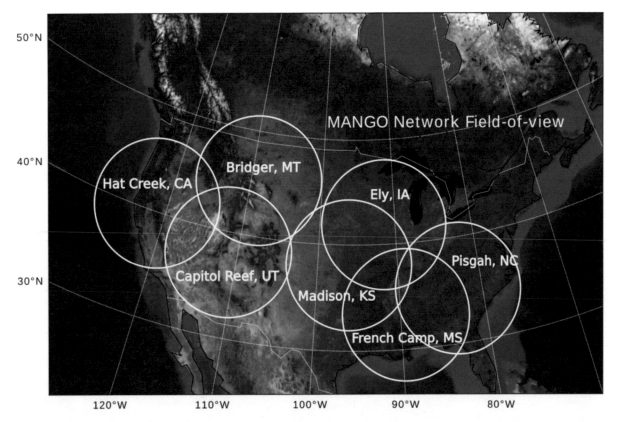

FIGURE 6-10 Midlatitude Allsky-imaging Network for GeoSpace Observations (MANGO) network-centric architecture for imaging large-scale airglow and aurora features.
SOURCES: Phil Erickson, Massachusetts Institute of Technology Haystack Observatory, presentation to workshop, April 14, 2022, from MANGO Project Team, SRI International, used with permission.

As measurements are being gathered for a particular purpose, their anticipated future use should be kept in mind when deciding on which metadata are being stored. For example, in the case of meta-instruments such as GNSS TEC[1] and MANGO (Figure 6-10), the "instrument" is the software used to combine the networked data collected for another purpose. However, Erickson pointed out that foreseeing what the future use of the data might be—and which metadata that might require—will be challenging.

Advances in space weather predictions will require multiscale observations with networks of sensors. Such operational space weather measurement systems must be designed as networks from the start. As discussed in other panels, the OSSEs can characterize optimal instrument placement, performance, and use, and thus help in designing the network. Furthermore, to make full use of the data will require incorporating the observations into data assimilation schemes.

---

[1] In the GNSS-TEC database, global total electron content (TEC) data were derived from data collected by a Global Navigation Satellite System (GNSS).

Finally, nontraditional data sources, such as the amateur radio community, can provide potentially useful observations with unique spatiotemporal sampling rates. However, the use of such crowd-sourced data has its own challenges related to quality and availability. The different needs of the end users may limit the use of such data sets—retrospective analysis for scientific advances can be more accommodating than the operationally oriented space weather applications, whose needs require a rigorous approach to data production.

Erickson also commented on the workforce development pressures that arise from the need to understand the production-level requirements of the space weather systems. This shared understanding can only be developed if the O2R and R2O efforts are done in interagency cooperation.

In her presentation, Building a Fluently Operating Architecture, Greer described how ionosphere–thermosphere–magnetosphere science could be addressed by heterogeneous data sets from ground- and space-based instrument platforms provided by governments, industry, and academia (Figure 6-11). Such an approach would require new approaches to overcome the challenges related to the calibration, validation, and verification needed to take advantage of the full power of large, distributed constellations. Cross-platform and cross-instrument comparisons should be part of the mission design from the start, especially when considering long-term monitoring. Data assimilation will be required to make full use of the measurements.

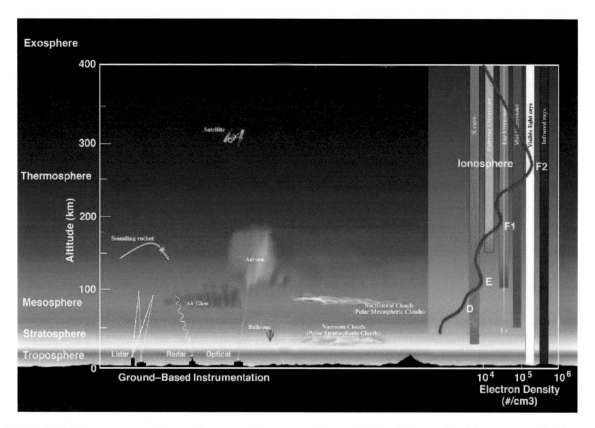

FIGURE 6-11 Heterogeneous data sets from a multi-source, multi-organizational observational base are needed for an operating architecture.
SOURCE: Katelynn Greer, University of Colorado Boulder, presentation to workshop, April 14, 2022.

Greer also noted the need for resources and agency coordination to ensure that instruments will be available and operational on the timelines required and that the instrument operators are incentivized to support the data taking. While this may not be considered the most glamorous work, it will be necessary for successful use of distributed resources.

Millan's presentation, The Radiation Belt Environment, emphasized how the infrastructure for the new space applications increases the number of satellites that encounter the radiation belts and their effects. Specific elements of the fleet include GPS, weather, and communications satellites. The space weather effects that the spacecraft encounter can be either localized or extended across the system; the strongest particle fluxes—and thus the greatest hazards to spacecraft—occur in the region bounded by the geosynchronous orbit (GEO) from above low Earth orbit (LEO) from below.

The end of NASA's Van Allen Probes mission in 2019 meant losing continuous monitoring of the radiation environment from the geostationary transfer orbit, which cuts through the critical radiation belt region. Those measurements cannot be replaced by either the data from the Geostationary Operational Environmental Satellite (GOES) system at GEO orbit, where the fluxes are much lower (Figure 6-12), or by operational data from the LEO orbits, which sample only a small portion of the particle distribution. Continuous monitoring of the radiation belts would be important both for scientific understanding, as predictive capability does not yet exist, and for operational space weather monitoring purposes to protect assets in orbit.

Millan described ongoing efforts to take advantage of opportunities to gain radiation belt monitoring capability. Given sufficient advance planning and resources, real-time space weather beacons, built based on the experience from the Van Allen Probes, could be realized in collaborations between NASA, NOAA,

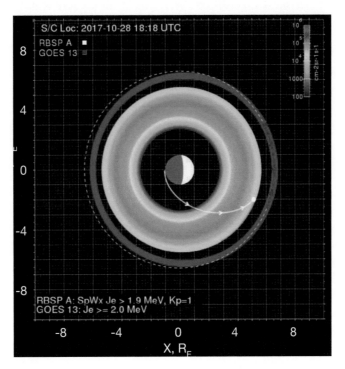

FIGURE 6-12 Radiation belt environment from the Radiation Belt Storm Probe (RBSP)-A and the Geostationary Operational Environmental Satellite (GOES) system.
SOURCE: Robyn Millan, Dartmouth College, presentation to workshop, April 14, 2022.

and international partners. Small satellites, such as CubeSats at LEO (built by the National Science Foundation, NASA, or international partners) or the Geostationary Transfer Orbit Satellite (GTOSat) flying through the high-radiation environment, offer new opportunities for radiation belt monitoring. Hosted payload approaches include the Responsive Environmental Assessment Commercially Hosted (REACH) constellation and those on Iridium NEXT, both of which are currently confined to LEO.

The collection of multi-point distributed measurements is a critical need for science, particularly as input into empirical and physics-based models. However, Millan said that it is not yet clear which measurements, which measurement locations, and how dense a network will be required to provide the scientific advances, as such studies have not been carried out in a comprehensive way.

Nowcasting and forecasting need real-time monitoring. Ensuring sufficient observational input for space weather products will require policies and programs that would increase the availability of measurements, including private–public partnerships, data buys, and addressing the barriers preventing international cooperation (an area the Committee on Space Research [COSPAR]) panel on space weather is seeking to address).

Furthermore, distributed observations present challenges that require new solutions. Onboard processing and artificial intelligence will be needed to address issues with data downlink and pipelines to operational centers. Space-based observations need to be combined with simulation, models, and ground-based data sources, which may require new approaches.

Finally, using commercial data has proven to be labor intensive and limited by the architecture that defines the resolution, availability, and other key metrics. Getting the best possible observations requires both coordination between the commercial and academic partners and buy-in from the commercial entities.

Impact of Commercialization on New Architectures for Space Weather

Anderson's presentation, Commercial Data Acquisition for Space Weather, focused on the various challenges of acquiring space weather data from commercial providers. One challenge is that U.S. government procurement processes and timelines are not well suited to engaging with commercial providers, because commercial entities move faster and have different mechanisms for approving commitment to projects. This leads to a situation in which the needs of scientific data collection are generally addressed only after the measurement system has been built.

Including custom instruments onboard commercial satellites requires close interaction with vendors, as support for instrument integration and operation may require significant vendor involvement. Generally, customization may not be possible, and the commercial vendors give no guarantees of performance versus requirements. Getting real-time data from a commercial partner requires a value proposition that appeals to the provider—opportunities for data buys arise once the product is developed. Anderson said that there are significant differences between the data acquisition strategies for science use and real-time space weather use: For research, data acquisition can be economical since delays and gaps can be tolerated. For space weather, all data have to be available in real time, and processing needs to be robust.

Low Earth orbit hosts an increasing number of sensors, which form a heterogeneous network that offers opportunities both for scientific and space weather use. Data acquisition for research can be economical, particularly if it is done on a non-interference basis with a commercial mission without assurance of continuous data access. However, scientific use typically requires intercalibration of the sensors, and data assimilation into a common framework can be a challenge if there are data quality issues or if required ancillary data are not readily available. Space weather use poses more requirements to the provider. Real-time data access may affect various system and staffing issues, as on-call staffing must be robust and available 24/7. In such cases the benefit to the commercial provider must be strong.

Anderson concluded that commercial data tend to be heterogeneous. The distribution of observations, although it may be "global," is dynamic and uneven. There may be very different levels of accuracy and noise in the measurements, so the errors and uncertainties of each data source need to be quantified, and covariance verification should be used to cast the information in a useful form for assimilation analyses. Such problems can be treated with multi-parameter assimilation, but even if it has been used for tropospheric weather, it is still an emerging capability and would require significant new research before it could be applied to operational products. Nevertheless, such methods may provide new insights for first-principles models for intense and unusual events.

In the panel's final presentation, Space Weather and Space Situational Analysis, Babcock, from Starlink Guidance, Navigation, and Control at SpaceX, provided a commercial perspective on the importance of space weather data for constellation management and collision avoidance in low Earth orbit. Both tasks require a good understanding of space weather processes, most importantly the ability to predict thermospheric density. While Babcock indicated that SpaceX is willing to share data with the scientific community, he noted the overhead associated with data production and commented that SpaceX is working to determine its impacts.

SpaceX has a large number of vehicles sampling relevant parameters in low Earth orbit, but it is not clear how to best gather data from the constellation. Babcock saw data gathering from the Starlink platforms as a potential game changer for in-space connectivity for science missions. If the laser terminals would be available to vehicles, that would open the possibility for continuous connectivity from an existing commercial provider.

Interactive Panel Discussion

Baker, the panel chair, reflecting a personal view that he believed was also the sense of participants, opened the discussion with the question: Have we made it clear to policy makers that we need a complete, robust, perpetual space weather system, and how the future architectures will support this need?

Two themes arising during discussions of the space weather infrastructure were the adequacy of resources and the need to plan ahead. Millan asked Babcock about the possibility of using student internships or postdocs to connect academia with the commercial community. Babcock responded that this might be a possible approach, but commented that a major issue in working with data from any commercial entity is their proprietary nature. For SpaceX, he said, the questions are which data will they be comfortable sharing, and how can they help the community while minimizing exposure to SpaceX and Starlink systems. The company already has plans about data delivery and frequency, and thoughts about activities going forward that would be beneficial for the community. However, in the discussion it was not clear who would carry out the engineering necessary to enable seamless commercial data sharing.

Greer asked what would incentivize SpaceX to become more engaged in obtaining the data concerning collision avoidance that Babcock indicated would be of interest. An ideal outcome would be a new, more precise, density model that the company could relay up to satellites for on-board calculations to support autonomous collision-avoidance maneuvers. Having the calculations done on board is necessary to minimize latency by eliminating the delay associated with the ground communications—density data should be updated as frequently and as quickly as possible. Minimizing the overall latency in the collision-avoidance maneuver process is a particular concern now as the solar maximum is approaching and the low Earth orbit is becoming more densely populated. This will mean minimizing latency in getting density updates and also in sharing data to update the state of debris.

On the topic of commercial-to-scientific engagement, Babcock said, SpaceX would prefer to establish a partner—such as NASA or NOAA—that would agree to host and publish data that SpaceX shares for public consumption, which would limit the interface with the scientific community to a single partner hosting the data.

Baker raised the question of what part of the operational space weather system will be available in the near-Earth environment, and what will be needed in the future beyond the research assets available for operational space weather systems today. A consistent theme throughout the workshop was the need to increase the availability and use of the already existing data for space weather research and operations. Work needs to be done to ensure that the existing data are usable, and that the new technologies are assessed and, where well-suited, incorporated into the new architectures. Furthermore, it is important to identify the core capabilities that are needed in space and on ground to feed the models. It was suggested that a workshop on this topic alone would be of great value.

Moving toward an operational environment will require that good careers exist in the area to attract the people who will work there, Anderson observed.

Baker then focused on preparing new observational platforms for the upcoming solar maximum. Is there an urgency to get new platforms flying, or is the community comfortable with what is already available? It was noted that there are some potential deficiencies in both engineering and operational categories. Babcock said that from the SpaceX perspective there are two areas of concern: There are engineering challenges in ensuring that designs take into account low-probability transient events, such as extreme geomagnetic storm events, as typically the designs have not accounted for situations that occur only 1 percent of the time. Operationally, the challenge is collision avoidance, and SpaceX is comfortable with the current risk levels. Nevertheless, Babcock added, there is an urgency regarding the upcoming solar maximum, particularly in light of the recent SpaceX Starlink losses, suggesting that the biggest gap in understanding is how to include the effects of rare, short-lived, but strong transients that are not captured in the models used in the engineering design phase, and therefore can lead to unpredicted consequences. These can, potentially, become of greater importance in the design of future architectures.

Greer spoke about the need for calibration and validation activities and the importance of continued support for the operations of instruments providing data, all of which are critically important as enablers of the use of heterogeneous data sources. The effort going into calibration and validation activities changes during the mission lifetime, but the activities can be made more efficient with good advance planning.

Regarding world-class facilities focused on discovery science, the discussion turned to the question of whether there is a way to establish baseline agency funding for their long-term support (for space weather purposes). It is important not to assume that the capabilities (in particular instruments) in place today will remain intact in the future. Erickson observed that the nature of space weather activities requires an explicit effort to include fundamental exploratory work that leads to new discoveries and innovations. Norton and Baker suggested that public–private partnerships could be used to generate data products that satisfy both science and operational requirements, and that this could be useful in sustaining infrastructure and capabilities for both space weather operations and scientific research.

Tomoko Matsuo suggested that a potential way to establish connections between academia and industry would be to have student projects focusing on space weather questions relevant to companies. The projects would benefit the students who get exposed to realistic problems as well as the companies who get answers to their pressing problems.

Paxton said that when commercial entities are involved, space weather needs to be thought of in terms of return on investment. For example, Iridium's support for the AMPERE (Active Magnetosphere and Planetary Electrodynamics Response Experiment) is driven by payment for the data that its satellites collect. Thus, companies might be persuaded to share their data (e.g., on satellite drag) if the data were used to create a product that the companies would benefit from. Switching gears, Paxton argued that space weather should also be thought about as a discovery activity. "There are so many things we should discover," he said, continuing that he would like a message to emerge from the workshop that there is fundamental science to be done, and not just improving existing models to predict various aspects of space weather.

Mary Hudson asked the panel whether they were satisfied with existing near-term measurement capabilities, particularly given that the Van Allen Probes and the Defense Meteorological Satellite Program (DMSP) are being phased out. Millan answered that the situation is not satisfactory, noting that given the long development times, it will likely not be until after the next solar maximum that there are new systems that can replace the DMSP or the Van Allen Probes observations. Several participants also pointed out the need for multi-point measurements at the higher altitudes above the radiation belts, especially in the nightside magnetotail.

A recurrent theme in the presentations and discussions was the need for multi-scale measurements. As space storms involve processes that occur in multiple scales, improving space weather predictions will require understanding how those scales operate and couple together.

Finally, the variability of the space weather impacts in relation to storm size using traditional metrics was discussed in relation to the February 4, 2022, storm. Anderson pointed out how the loss of the StarLink satellites is an example of a small storm that had significant space weather impacts—an issue that needs to be understood and addressed.

REFERENCE

Desai, M.I., and D. Burgess. 2008. "Particle Acceleration at Coronal Mass Ejection–Driven Interplanetary Shocks and the Earth's Bow Shock." *Journal of Geophysical Research* 113:A00B06.

Appendixes

A

Statement of Task

The National Academies of Sciences, Engineering, and Medicine will appoint an ad hoc committee to organize a workshop that will focus on the research agenda and observations needed to improve the understanding of the Sun–Earth system that generates space weather consequences. Specifically, the Phase II workshop will:

- Examine trends in available and anticipated observations, including the use of constellations of small satellites, hosted payloads, ground-based systems, international collaborations, and data buys, that are likely to drive future space weather architectures; review existing and developing technologies for both research and observations;
- Consider the adequacy and uses of existing relevant programs across the agencies, including NASA's Living With a Star (LWS) program and its Space Weather Science Application initiative, the National Science Foundation's (NSF's) Geospace research programs, and NOAA's Research to Operations (R2O) and Operations to Research (O2R) programs for reaching the goals described above;
- Consider needs, gaps, and opportunities in space weather modeling and validation, including a review of the status of data assimilation and ensemble approaches;
- Consider how to incorporate data from NASA missions that are "one-off" or otherwise non-operational into operational environments, and assess the value and need for real-time data (for example, by providing "beacons" on NASA research missions) to improve forecasting models; and
- Take into account the results of studies, including NASA's space weather science gap analysis (part of the NASA Heliophysics Division's Space Weather Science Application program) and the NSF *Investments in Critical Capabilities for Geospace Science* (2016), to identify the key elements needed to establish a robust research infrastructure.

A proceedings summarizing the presentations and discussions at the workshop will be prepared by the committee in accordance with institutional guidelines. The workshop proceedings will not include findings or recommendations.

B

Workshop Agenda

Public Agenda—Virtual
April 11–14, 2022

All Times Are EDT

DAY 1: MONDAY, APRIL 11, 2022

11:00 am   Keynote on Results from Phase I Workshop and Proceedings
<u>Moderator</u>: *Tuija Pulkkinen,* Workshop Organizing Committee Co-Chair

Mary Hudson, Dartmouth College
Janet Luhmann, University of California, Berkeley

11:30 am   Agency Panel: Recent Updates
<u>Moderator</u>: *Mary Hudson,* Committee Member

Jim Spann, NASA
Mangala Sharma, National Science Foundation (NSF)
Elsayed Talaat, National Oceanic and Atmospheric Administration (NOAA)

12:15 pm   Break

1:00 pm   Interagency Partnerships Panel: New Ways of Working
<u>Moderator</u>: *Tuija Pulkkinen,* Workshop Organizing Committee Co-Chair

Jinni Meehan, NOAA/National Weather Service
Dan Moses, NASA/HPD
Tammy Dickinson, Science Matters, Inc.
Sage Andorka, U.S. Space Force

1:45 pm   Space Weather Operations Panel
               <u>Moderator</u>: *Delores Knipp,* Committee Member

               Mark Olson, NERC
               Hazel Bain, NOAA/SWPC
               Michele Cash, NOAA/SWPC
               Mike Stills, Villanova University
               Scott Leonard, NOAA/OSC

2:15 pm   People Perspective Panel
               <u>Moderator</u>: *Nicki Viall,* Committee Member

               Eddie Gonzalez, NASA Goddard Space Flight Center
               MacArthur Jones, NRL
               Fran Bagenal, University of Colorado

2:45 pm   Break

3:30 pm   Keynote Presentation: New Research Needs
               <u>Moderator</u>: *Christina Cohen,* Workshop Organizing Committee Co-Chair

               Drew Turner, JHU/APL
               Judy Karpen, NASA/GSFC
               Noé Lugaz, University of New Hampshire
               Katie Whitman, NASA/JSC

4:50 pm   Keynote Presentation: Prediction of Ground Effects
               <u>Moderator</u>: *Christina Cohen,* Workshop Organizing Committee Co-Chair

               Jeff Love, USGS

5:20 pm   Adjourn Open Session

DAY 2: TUESDAY, APRIL 12, 2022

11:00 am   Ionospheric State and Irregularities Panel
               <u>Moderator</u>: *Anthea Coster,* Committee Member

               Seebany Datta-Barua, Illinois Institute of Technology
               Charles Carrano, Boston College
               Jonathan Snively, Embry-Riddle Aeronautical University
               Sean Bruinsma, CNES
               Greg Ginet, MIT

11:40 am   Cross-Scale and Cross-Region Coupling Panel
               <u>Moderator</u>: *Endawoke Yizengaw,* Committee Member

               Josh Semeter, Boston University
               Jonathan Rae, Northumbria University, Newcastle upon Tyne, England

Astrid Maute, UCAR
Joe Huba, Syntek Technologies, Inc.
Seth Claudepierre, UCLA

12:20 pm   Break

1:00 pm   Solar Panel
                <u>Moderator</u>: *Pete Riley,* Committee Member

                Todd Hoeksema, Stanford University
                Sarah Gibson, UCAR
                Cooper Downs, Predictive Science Inc.
                Craig DeForest, Southwest Research Institute
                Valentin Pillet, National Solar Observatory
                Phil Chamberlin, University of Colorado

1:45 pm   Solar Wind Panel
                <u>Moderator</u>: *Nicki Viall,* Committee Member

                Joe Borovsky, Space Science Institute
                Vic Pizzo, NOAA/SWPC
                Stuart Bale, UC Berkeley
                Nick Arge, NASA/GSFC
                Erika Palmerio, Predictive Science Inc.

2:30 pm   Magnetosphere Panel
                <u>Moderator</u>: *Terry Onsager,* Committee Member

                Geoff Reeves, Los Alamos National Laboratory
                Christine Gabrielese, UCLA
                Larry Kepko, NASA/GSFC
                Lauren Blum, University of Colorado
                Matina Gkioulidou, JHU/APL
                Vania Jordanova, Los Alamos National Laboratory

3:15 pm   Break

4:00 pm   Ionosphere and Thermosphere Panel
                <u>Moderator</u>: *Larry Paxton,* Committee Member

                Naomi Maruyama, University of Colorado
                Matt Zettergren, Embry-Riddle Aeronautical University
                Katrina Bossert, Arizona State University
                Bill Lotko, UCAR
                Larisa Goncharenko, MIT
                Hanli Liu, UCAR

4:45 pm   Ground Effects Panel
                 <u>Moderator</u>: *Delores Knipp*, Committee Member

                 Adam Schultz, Oregon State University
                 Jenn Gannon, Computational Physics Inc.
                 Jesper Gjerloev, JHU/APL
                 Arnaud Chulliat, University of Colorado, NOAA/NCEI
                 Antti Pulkkinen, NASA/GSFC
                 Anna Kelbert, USGS

5:30 pm   Adjourn Open Session

DAY 3: WEDNESDAY, APRIL 13, 2022

11:00 am   Data Science and Analytics: Keynotes
                 <u>Moderator</u>: *Delores Knipp*, Committee Member

                 Ricardo Todling, NASA/GSFC
                 Enrico Camporeale, CU/CIRES

11:45 am   Data Science and Analytics: Data/Model Resources and Curation Panel
                 <u>Moderator</u>: *Anthea Coster*, Committee Member

                 Carrie Black, NSF
                 William Schreiner, UCAR
                 Jack Ireland, NASA
                 Rob Redmon, NOAA
                 Masha Kuznetsova, NASA/CCMC
                 Alec Engell, NextGen Federal Systems

12:30 pm   Break

1:15 pm   Data Science and Analytics: Data Fusion and Assimilation Panel
                 <u>Moderator</u>: *Charles Norton*, Committee Member

                 Tomoko Matsuo, University of Colorado
                 Alex Chartier, JHU/APL
                 Bernie Jackson, UCSD
                 Mark Cheung, Lockheed-Martin
                 Eric Blasch, Air Force Office of Scientific Research
                 Slava Merkin, JHU/APL

2:00 pm   Data Science and Analytics: Machine Learning and Validation Panel
                 <u>Moderator</u>: *KD Leka*, Committee Member

                 Jacob Bortnik, UCLA
                 Asti Bhatt, SRI

Shasha Zou, University of Michigan
Morris Cohen, Georgia Institute of Technology
David Fouhey, University of Michigan
Hannah Marlowe, Amazon Web Services

2:45 pm   Break

3:30 pm   New Architectures: Solar and Heliosphere Panel
               <u>Moderator:</u> *Dan Baker,* Committee Member

               Justin Kasper, BWX Technologies
               Nicole Duncan, Ball Aerospace
               Tom Berger, University of Colorado
               Sue Lepri, University of Michigan
               Angelos Vourlidas, JHU/APL

5:00 pm   Poster Session
               <u>Moderator:</u> *Ron Turner,* Committee Member

6:00 pm   Adjourn Open Session

DAY 4: THURSDAY, APRIL 14, 2022

11:00 am   Data Science and Analytics: Ensemble Modeling Panel
               <u>Moderator:</u> *Mary Hudson,* Committee Member

               Eric Adamson, NOAA/SWPC
               Kent Tobiska, Spacewx SET
               Dan Welling, University of Texas Arlington
               Sean Elvidge, University of Birmingham
               Nick Pedatella, UCAR
               Edmund Henley, UK Met Office

11:45 am   New Architectures: Magnetosphere–Ionosphere–Thermosphere Panel
               <u>Moderator:</u> *Dan Baker,* Committee Member

               Phil Erickson, MIT
               Brian Anderson, JHU/APL
               Allison Jaynes, University of Iowa
               Katelynn Greer, University of Colorado
               Mike Nicolls, SpaceX

1:15 pm   Adjourn Open Session

2:00 pm   Committee Closed Session

TBD   Meeting Adjourns NLT 5:00 pm

C

Poster Session at the
April 11–14, 2022, Workshop

The workshop included a poster session to provide the community with an opportunity to discuss topics relevant to the workshop (research needs, capabilities, data infrastructure/analytics, and machine learning). Posters were introduced in a pre-recorded, virtual "Lightning Round" (3 minutes per poster provided by the presenter). Links to the posters and to the video presentations were available to attendees throughout the workshop and are also included online.[1] Thirteen posters were submitted. The title, author, links, and summary information about each of the posters are provided below, grouped into four, sometimes overlapping, categories.

- Solar and solar wind (3)
- Magnetosphere, ionosphere, and thermosphere (3)
- Data science, analytics, and ensemble modeling (3)
- Machine learning (4)

SOLAR AND SOLAR WIND

Moon to Mars (M2M) Space Weather Analysis Office
Yaireska (Yari) Collado-Vega, NASA GSFC
https://vimeo.com/showcase/9407816/video/699034448
The presentation provided a description of the new NASA Moon to Mars (M2M) Space Weather Analysis Office mission and goals. The Office was established to test novel capabilities in collaboration with CCMC and the Space Radiation Analysis Group at Johnson Space Center, serving as the proving ground of real-time analysis to characterize the radiation environment (including both lunar and Mars missions). M2M also supports NASA robotic missions and, by serving as a proving ground, its capabilities

---

[1] Links to the posters and videos can be found at National Academies of Sciences, Engineering, and Medicine, "Space Weather Operations and Research Infrastructure: Proceedings of the Phase II - (Workshop)," https://www.nationalacademies.org/event/04-11-2022/space-weather-operations-and-research-infrastructure-workshop-phase-ii-workshop.

can be used for commercial purposes if transitioned to operational agencies. M2M analysts also perform real-time validation after solar events and characterize the limitations of the models used in forecasting.

Space Weather at Mars: Research (and Forecasting) Needs

Christina Lee, Space Sciences Laboratory, UC Berkeley

https://vimeo.com/showcase/9407816/video/697514529

Space weather research and forecast needs at Mars were discussed. MARSIS (Mars Advanced Radar for Subsurface and Ionosphere Sounding), for example, experienced a 10-day blackout from SEPs in September 2017. There is a history of space weather observations, where MAVEN (Mars Atmosphere and Volatile Evolution mission) has the most complete coverage, but data gaps occur frequently and sometimes for long durations. Continuous upstream measurements, like an ACE at Mars L1, are needed. Continuation of new instrument payloads on future missions would be valuable. Currently, forecasting at Mars requires real-time beacons from Earth. This is a problem during conjunction, so a dedicated space weather monitor will be needed for Mars inhabitants.

New Developments in Space Weather Modeling

Gabor Toth, University of Michigan

https://vimeo.com/showcase/9407816/video/697513568

The poster described an open-source model of background solar wind with uncertainty quantification. Work continues on CME models to produce the best results in terms of arrival time and other characteristics. Current activity includes incorporating data assimilation for geospace modeling to improve CME prediction. Solar wind monitors would improve solar wind and CME forecasts, including L5 in situ observations and multi-point real-time white light images.

MAGNETOSPHERE, IONOSPHERE, AND THERMOSPHERE

3-D Regional Ionosphere Imaging and SED Reconstruction with a New TEC-Based Ionospheric Data Assimilation System (TIDAS)

Ercha Aa, MIT Haystack Observatory

https://app.smartsheet.com/b/publish?EQBCT=32396a6d7b4d4d85b61d72979abfc257

A new total electron content-based ionospheric data assimilation system (TIDAS) over the Continental U.S. is developed using a hybrid Ensemble-Variational scheme. This data assimilation system can provide accurate and reliable three-dimensional time-evolving electron density maps with high spatial-temporal resolution ($1° \times 1° \times 20$ km $\times 5$ min). This high-fidelity regional data assimilation system is a powerful space weather nowcasting tool to reconstruct localized storm-time ionospheric morphology with unprecedented and fine-scale details. Results can help advance current understanding of the fine structures and underlying mechanisms of the midlatitude ionospheric density gradients.

Exploring Coverage of Accelerometer Satellites in the Thermosphere: An Observing System Simulation Experiment

Eric Sutton, CU-Boulder/SWx TREC

https://vimeo.com/showcase/9407816/video/697513206

The poster described an Observing System Simulation Experiment assessing optimal satellite distribution in the thermosphere (8 satellites) for neutral density measurements. It was used to investigate the impact of satellite coverage. While it focused on a specific instrument, it can be expanded to other areas of exploration.

Measuring Upper Atmospheric Winds with the Zephyr Meteor Radar Network and a Look to the Future
Ryan Volz, MIT Haystack Observatory
https://app.smartsheet.com/b/publish?EQBCT=32396a6d7b4d4d85b61d72979abfc257

> This project addresses a significant gap in measuring mesospheric and lower thermospheric neutral winds. The presentation describes the NSF DASI (Distributed Array of Small Instruments) project to develop and build a meteor radar network near Boulder, CO, for estimating upper atmosphere neutral winds. A signal reflected from ionized meteor trails is used to estimate zonal and meridional wind with mesoscale horizontal resolution (10-50 km) as a function of altitude (80-100 km) and time. The results are strongly data-driven with minimal assumptions. Rigorous output uncertainties are derived from prior confidence, measurement errors, and sampling density. Covering the United States would require on the order of mid-scale funding. Such a system would provide insight into the lower atmosphere forcing of the upper atmosphere, using technology already proven in Germany.

DATA SCIENCE, ANALYTICS, AND ENSEMBLE MODELING

Geospace Data Systems Infrastructure: Current Needs and Status
Tai Yin Huang, National Science Foundation (NSF)
https://vimeo.com/showcase/9407816/video/694079474

> There are many unmet data infrastructure needs. NSF has many solicitations to support data infrastructure. Challenges at NSF in this area were also highlighted. Training courses to how to use data resources and to improve open access is recommended. Collaboration with federal agencies, a community workshop, and other preliminary findings are recommended to address unmet infrastructure needs.

The "Silent" Technology We Need: The Earth and Space Science Knowledge Commons
Ryan McGranaghan, Orion Space Solutions LLC; NASA Goddard Space Flight Center
https://vimeo.com/showcase/9407816/video/694080757

> What emerging topic(s) are we missing? This poster describes the idea of the "Knowledge Commons," a combination of intelligent information representation and the openness, governance, and trust required to create a participatory ecosystem whereby the whole community maintains and evolves this shared information space. In describing the history of knowledge graphs and commons across contexts and nascent work to create them for heliophysics, we identify the technologies that are prerequisite to more robust, responsive, and responsible data assimilation, machine learning, and ensemble modeling. In organizing thoughts to guide the committee and community around these underlying technologies, we arrive at very concrete recommendations for the future of space weather operations and research and for our increasingly wide-reaching community.

Improving Predictive Skill with Ensemble Modeling
Steven K. Morley, Los Alamos National Laboratory
https://vimeo.com/showcase/9407816/video/694080162

> This presentation noted that it is critical to capture uncertainty and to propagate it through simulations. The SWMF and Minimal Substorm Model (MSM) were related to geometrical hazards. Ensemble modeling allows greater fidelity than single-forecast models where physical models can be outperformed. Ensemble modeling is tractable and can be achieved at modest computational cost, and can improve the skill of predictions

MACHINE LEARNING

Convection Patterns Based on Machine Learning

William Bristow, Penn State University

https://app.smartsheet.com/b/publish?EQBCT=32396a6d7b4d4d85b61d72979abfc257

The poster describes a study that combines data from the Super Dual Auroral Radar Network (Super-DARN) with a machine-learning (ML) model of convection based on five years of SuperDARN observations. The model is driven with a set of parameters: measures of the magnetospheric state (indices), solar wind/IMF parameters and drift vectors resolved into N-S and E-W, in each latitude-MLT grid cell. Three regressor algorithms provided by the Scikit-Learn software package were tested for forming the model. The Random Forest Regressor produced the lowest root mean square error. The ML model seems to develop a "memory" that responds to expansion and contraction of the polar cap in latitude in response to changing Al and Au indices.

Magnetoseismology for Space Weather Operations and Research

Peter Chi, University of California, Los Angeles

https://vimeo.com/showcase/9407816/video/697514066

Seismology is a well-established method. Ground-based networks can make these measurements. Plasma density profiles can be measured where travel-time magneto-seismology can give information about the magnetotail. Machine learning is needed to facilitate big data analysis.

Prediction of Near–Bow Shock Solar Wind Conditions

Terry Liu, University of California, Los Angeles

https://vimeo.com/showcase/9407816/video/694076888

Multipoint observations are needed to capture near–bow shock localized solar wind conditions. New concentric spacecraft constellations with combination of physics-based prediction models, assimilative reconstruction, and machine language including data from Magnetospheric Multiscale (MMS) mission and THEMIS can be used to train predictive models. This can lead to ten minute to one hour forecast warnings. Predicting near-bow shock solar wind is needed for this observation.

Machine Learning for Thermosphere Operations and Science

Piyush M. Mehta, West Virginia University

https://vimeo.com/showcase/9407816/video/694084508

The thermosphere is the largest source of uncertainty in LEO operations. Performance comparisons including during geomagnetic storms show that data-driven machine learning methodologies can help improve the overall fidelity, resolution, and accuracy of both empirical and physics-based models. They can help improve scientific understanding of processes and provide a capability for robust and reliable uncertainty quantification.

D

Acronyms and Abbreviations

ACE	Advanced Composition Explorer
ADAPT	Air Force Data Assimilative Photospheric Flux Transport
AI	artificial intelligence
AIP	American Institute of Physics
AMPERE	Active Magnetosphere and Planetary Electrodynamics Response Experiment
AMR	adaptive mesh refinement
ANSWERS	Advancing National Space Weather Expertise and Research toward Societal Resilience
AOM	Arctic Observing Mission
ARC	Applied Research Challenge
CCMC	Community Coordinated Modeling Center
CCOR	Compact Coronagraph
CEDAR	Coupling, Energetics, and Dynamics of Atmospheric Regions
CIR	corotating interaction region
CME	coronal mass ejection
COE	Center of Excellence
COSMIC-2	Constellation Observing System for Meteorology, Ionosphere, and Climate 2
COSPAR	Committee on Space Research
COTS	commercial off-the-shelf
CSA	Canadian Space Agency
CuSP	CubeSat for Solar Particles
CV	computer vision
DA	data assimilation
DMSP	Defense Meteorological Satellite Program

DoD   Department of Defense
DSCOVR   Deep Space Climate Observatory
DSN   NASA's Deep Space Network
Dst   disturbance storm time
EMP   electromagnetic pulse
EPP   energetic particle precipitation
ESA   European Space Agency
ESIP   Earth Science Information Partners
EUV   extreme ultraviolet
EZIE   Electrojet Zeeman Imaging Explorer

FAC   field-aligned current
FAIR   findable, accessible, interoperable, reusable

GDC   Geospace Dynamics Constellation
GEM   Geospace Environment Modeling
GEO   geosynchronous orbit
GEO/AGS   Directorate of Geosciences Division of Atmospheric and Geospace Sciences (NSF)
GIC   geomagnetically induced current
GMD   geomagnetic disturbance
GNSS   Global Navigation Satellite System
GOES   Geostationary Operational Environmental Satellite
Golf-NG   Global Oscillations at Low Frequency-New Generation
GONG   Global Oscillation Network Group
GPS   Global Positioning System
GPU   graphics processing unit
GTOSat   Geostationary Transfer Orbit Satellite

HASDM   High-Accuracy Satellite Drag Model
HDRL   Heliophysics Digital Resource Library (NASA)
HERMES   Heliophysics Environmental and Radiation Measurement Experiment Suite
HF   high frequency
HMI   Helioseismic and Magnetic Imager
HPD   Heliophysics Division
HSO   Heliophysics System Observatory
HSS   high-speed stream

ICME   interplanetary coronal mass ejection
IGRF   International Geomagnetic Reference Field
IMF   interplanetary magnetic field
IRIS   In Situ and Remote Ionospheric Sensing
ISR   incoherent scatter radar
ISS   International Space Station
ISWAT   International Space Weather Action Team
ITM   ionosphere–thermosphere–mesosphere
IWG   interagency working group

JEDI   Joint Effort for Data-Assimilation Integration
JH   joule heating
JHU/APL   Johns Hopkins University Applied Physics Laboratory

LASCO   Large Angle and Spectrometric Coronagraph Experiment
LEO   low Earth orbit
LWS   Living With a Star (NASA)

M2M   Moon to Mars
MANGO   Midlatitude Allsky-imaging Network for GeoSpace Observations
MEO   medium Earth orbit
MHD   magnetohydrodynamics
MIMIS   Mission to Investigate the Mesoscales in Interplanetary Space
ML   machine learning
MLT   mesosphere and lower thermosphere
MME   multi-model ensemble
MSL   Mars Science Laboratory

NASA   National Aeronautics and Space Administration
NCAR   National Center for Atmospheric Research
NCEI   National Centers for Environmental Information (NOAA)
NERC   North American Electric Reliability Corporation (NOAA)
NESDIS   National Environmental Satellite, Data, and Information Service
NOAA   National Oceanic and Atmospheric Administration
NSF   National Science Foundation
NSF/AGS   NSF Division of Atmospheric and Geospace Sciences
NSF/AST   NSF Division of Astronomical Sciences
NSTC   National Science and Technology Council
NSW-SAP   National Space Weather Strategy and Action Plan
NWS   National Weather Service

OADR   Open-Architecture Data Repository
OSE   Observing System Experiment
OSSE   observing system simulation experiment
OSTP   Office of Science and Technology Policy
OTHR   over-the-horizon radar

P&F   Pepperl & Fuchs
PIC   particle in cell
POES   Polar Operational Environmental Satellites
PROSWIFT   Promoting Research and Observations of Space Weather to Improve the Forecasting of
                Tomorrow Act
PUNCH   Polarimeter to Unify the Corona and Heliosphere

R2O2R   research to operations/operations to research
RAD   Radiation Assessment Detector

REACH   Responsive Environmental Assessment Commercially Hosted
RL   readiness level
ROI   return on investment
ROSES   Research Opportunities in Space and Earth Sciences

SBIR   Small Business Innovation Research
SDO   Solar Dynamics Observatory
SEP   solar energetic particle
SET4D   Space Domain Awareness Environmental Toolkit for Defense
SHINE   Solar, Heliospheric, and Interplanetary Environment
SIR   stream interaction region
SNIPE   Small Scale Magnespheric and Ionospheric Plasma Experiment
SOARS   Significant Opportunities in Atmospheric Research and Science
SOHO   Solar and Heliospheric Observatory
SPASE   Space Physics Archive Search and Extract
SpWx   space weather
SRAG   Space Radiation Analysis Group
SSA   space situational awareness
SSI   Synchronous Serial Interface
STEM   science, technology, engineering, and mathematics
STEREO   Solar Terrestrial Relations Observatory
SuperDARN   Super Dual Auroral Radar Network
SWAG   Space Weather Advisory Group
SWFO-L1   Space Weather Follow On-Lagrange 1
SWMF   Space Weather Modeling Framework
SWORM   Space Weather Operations, Research, and Mitigation
SWPC   Space Weather Prediction Center
SWPT   Space Weather Prediction Testbed
SWx   space weather
SWxSA   Space Weather Science Application Program

TEC   total electron content
THEMIS   Time History of Events and Macroscale Interactions during Substorms
TID   traveling ionospheric disturbance
TRL   technology readiness level

USGS   U.S. Geological Survey
UV   ultraviolet

WAM-IPE   Whole Atmosphere Model–Ionosphere Plasmasphere Exosphere
WIPSS   Worldwide Interplanetary Scintillation Stations
WMM   World Magnetic Model

E

Biographies of Committee Members and Staff

CHRISTINA M.S. COHEN, *Co-Chair*, is a member of the professional staff at the California Institute of Technology where her work involves the design, calibration, and analysis of several energetic particle instruments. Dr. Cohen's research currently focuses on the acceleration, transport, and properties of solar energetic particles in the heliosphere and their space weather implications; and previously has included energetic particle populations in the Jovian magnetosphere and the heavy ion composition of the solar wind. Dr. Cohen analyzes combined in situ particle measurements with remote sensing of flares, radio bursts, and coronal mass ejections. Dr. Cohen is the principal investigator (PI) on the Ultra-Low Energy Isotope Spectrometer (ULEIS) on the Advanced Composition Explorer (ACE) mission, as well as co-investigator on the High-energy Ion Telescope (HIT) on the Interstellar Mapping and Acceleration Probe (IMAP) mission, the Energetic Particle Instrument-High (EPI-Hi) on Parker Solar Probe, and the Sun Radio Interferometer Space Experiment (SunRISE; a small satellite). Dr. Cohen is also a team member of the Solar Isotope Spectrometer (SIS) on ACE, the Low Energy Telescope (LET) on the Solar Terrestrial Relations Observatory (STEREO) mission, and the Heavy Ion Counter (HIC) on the Galileo mission. Dr. Cohen is on the science advisory board for *Eos* and is the past president of the space physics and aeronomy section of the American Geophysical Union (AGU). Dr. Cohen earned a Ph.D. in physics from the University of Maryland, College Park.

TUIJA I. PULKKINEN, *Co-Chair*, is chair and professor at the University of Michigan in Ann Arbor in the Department of Climate and Space Sciences and Engineering. Dr. Pulkkinen previously served as professor, vice president, and dean of the School of Electrical Engineering at the Aalto University in Espoo, Finland. Dr. Pulkkinen's research interests comprise Sun–Earth connection physics in a wide sense: energy transfer processes from the solar wind to the magnetosphere–ionosphere system; effects of large solar disturbances in the magnetosphere and ionosphere; auroral processes and their relationship to magnetotail dynamics; storm and substorm effects in the magnetotail, in the inner magnetosphere, and in the ionosphere; long-term solar variability effects in the geoefficiency of the solar wind driving; and space weather effects of solar wind–driven magnetospheric dynamics. Dr. Pulkkinen is an expert in empirical modeling of the magnetospheric magnetic field and in the development of quantitative analysis methods for global magnetohydrodynamic (MHD) simulations and in multi-instrument data analysis using measurements from space- and ground-based instruments. Dr. Pulkkinen is a member of the U.S. National Academy of Sciences and was the 2019 Birkeland Lecturer, The Norwegian Academy of Science and Letters, and a recipient in 2017 of the Julius Bartels Medal, European Geosciences Union. Dr. Pulkkinen received a Ph.D. in theoretical physics from the University of Helsinki.

DANIEL N. BAKER is director of the Laboratory for Atmospheric and Space Physics; Distinguished Professor of Planetary and Space Physics; and the Moog-Broad Reach Endowed Chair of Space Sciences, University of Colorado Boulder. Previously, Dr. Baker was group leader for Space Plasma Physics at Los Alamos National Laboratory and division chief at NASA's Goddard Space Flight Center. Dr. Baker's research is in spacecraft instrumental design and calibration, space physics data analysis, and magnetospheric modeling. Dr. Baker has studied plasma physical and energetic particle phenomena in the magnetospheres of Jupiter and Mercury, and the plasma sheet and magnetopause boundary regions of Earth's magnetosphere. Dr. Baker is experienced in the analysis of large data sets from spacecraft at geostationary orbit, and involvement in missions to Earth's deep magnetotail and comets, in the study of solar wind–magnetospheric energy coupling, and theoretical modeling of the possible role of heavy ions in the development of magnetotail instabilities. Dr. Baker's current interests include the use of computer systems and networks to enhance the acquisition, dissemination, and display of spacecraft data. Dr. Baker was lead investigator on several NASA space missions, including Magnetospheric Multiscale and the Radiation Belt Storm Probes (Van Allen Probes). Dr. Baker's honors include being awarded the 2018 William Bowie Medal from the AGU for outstanding geoscience research. Dr. Baker is a member of the National Academy of Engineering and was the recipient of the 2019 Hannes Alfvén Medal of the European Geosciences Union. Dr. Baker received his Ph.D. in space physics from the University of Iowa.

ANTHEA J. COSTER is a principal research scientist at the Massachusetts Institute of Technology's Haystack Observatory in Westford, Massachusetts. Dr. Coster's research interests include physics of the ionosphere, magnetosphere, and thermosphere; space weather and geomagnetic storm time effects; coupling between the lower and upper atmosphere; GPS positioning and measurement accuracy; radio wave propagation effects; and meteor detection and analysis. Dr. Coster is a co-PI on the National Science Foundation (NSF)-supported Millstone Hill Geospace facility award and a PI/co-PI on numerous projects involving the use of GPS to probe the atmosphere, including investigations of the plasma-spheric boundary layer, stratospheric warming, and the ionosphere over the Antarctic. Dr. Coster and co-workers developed the first real-time ionospheric monitoring system based on GPS. Dr. Coster has been involved with measuring atmospheric disturbances over short baselines (GPS networks smaller than 100 km) for the U.S. Federal Aviation Administration and has coordinated meteor research using the ALTAIR dual-frequency radar for NASA. Dr. Coster received a Ph.D. in space physics and astronomy from Rice University.

MARY K. HUDSON is the Eleanor and A. Kelvin Smith Professor Emerita of Physics at Dartmouth College and a senior research associate at the National Center for Atmospheric Research (NCAR). Dr. Hudson also served for 8 years as chair of the Department of Physics and Astronomy at Dartmouth. Current areas of investigation include the evolution of the radiation belts; how the ionized particle outflow is known as the solar wind and the magnetic field of the Sun interact with the magnetic field of Earth, producing electrical currents in the ionosphere; and the effects of solar cosmic rays on radio communications near Earth's poles. Dr. Hudson is a co-investigator on NASA's Van Allen Probes Mission and was one of the PIs with the Center for Integrated Space Weather Modeling, where researchers studied the weather patterns that originate from a solar eruption, following the energy and mass transfer through the interplanetary medium, all the way to Earth's ionosphere. Dr. Hudson is a fellow of the AGU, recipient of the 2017 Fleming Medal and the AGU Macelwane Award, and has served on the Heliophysics Subcommittee of the NASA Advisory Council. Dr. Hudson received a Ph.D. in physics from the University of California, Los Angeles.

DELORES KNIPP is a research professor in the Smead Aerospace Engineering Science Department and a senior research associate at the NCAR High Altitude Observatory. Dr. Knipp's research focuses on weather at the space–atmosphere interaction region; she also advances scientific use of space environment observations and promotes education related to space weather. Dr. Knipp has over 30 years of research and teaching experience in meteorology and upper atmosphere and geospace environment physics. Dr. Knipp is a retired U.S. Air Force officer, an American Meteorological Society fellow, and the director of the Education Enterprise at the University of Colorado Boulder's Space Weather Technology, Research, and Education Center. Dr. Knipp earned a Ph.D. in atmospheric science at the University of California, Los Angeles.

KD LEKA is a senior research scientist at the NorthWest Research Associates' Boulder, Colorado, office, and a designated foreign professor at the Institute for Space–Earth Environmental Research at Nagoya University, in Nagoya, Japan. Dr. Leka held post-doctoral fellowships at the National Oceanic and Atmospheric Administration (NOAA)/Space Environment Center through the National Research Council, and at the High Altitude Observatory through the NCAR/Advanced Study Program. Dr. Leka's present research interests span solar and space physics, including solar active regions, spectropolarimetry and magnetic fields, and solar energetic event prediction. Dr. Leka has served as chair of the User's Committee of the National Solar Observatories and on a NASA Senior Review of Heliophysics Operating Missions. Dr. Leka earned a Ph.D. in astronomy from the University of Hawaii.

CHARLES D. NORTON is the associate chief technologist at the NASA Jet Propulsion Laboratory (JPL) contributing to JPL's strategic planning, investment portfolio, new technology research, and infusion into flight projects. Dr. Norton recently served as the special advisor for small spacecraft missions at NASA Headquarters. While at NASA Headquarters, Dr. Norton was responsible for advising on cross-agency strategic directions for innovative small satellite science, exploration, and technology missions—from ESPA-Class spacecraft down to CubeSats. Dr. Norton has led and performed research spanning high-performance computing, advanced information systems technology, and small satellite science and technology mission development. Additionally, Dr. Norton has managed multiple CubeSat flight projects for NASA and co-led the Caltech KISS Study Program, "Small Satellites: A Revolution in Space Science." Dr. Norton is a recipient of numerous awards for new technology and innovation, including the JPL Lew Allen Award, the Voyager Award, and the NASA Exceptional Service Medal. Dr. Norton received a Ph.D. in computer science from Rensselaer Polytechnic Institute.

TERRANCE G. ONSAGER is a physicist with the NOAA Space Weather Prediction Center. Dr. Onsager's research includes solar wind–magnetosphere coupling, modeling the signatures of magnetic reconnection at Earth's magnetopause and in the magnetotail, and the dynamics of the electron radiation belts. It also includes coordinating the capabilities and priorities of international space weather organizations to improve global space weather services and working to bridge the gap between research and operations. Dr. Onsager has served as director of the International Space Environment Service, as co-chair of the World Meteorological Organization Inter-Programme Coordination Team on Space Weather, and as a member of the Space Weather Expert Team for the United Nations Committee on the Peaceful Use of Outer Space Working Group on the Long-Term Sustainability of Outer Space. Dr. Onsager received a Ph.D. in physics from the University of Washington with a focus on shock waves in collisionless plasma, using Earth's bow shock as a natural laboratory.

LARRY J. PAXTON is a member of the principal professional staff at the Johns Hopkins University Applied Physics Laboratory (JHU/APL) and is chief scientist for Geospace. His research interests include space science, space technology, satellite- and ground-based mission design, the implications of global climate change for the stability of nations, and innovation. Dr. Paxton is particularly interested in new instruments that characterize the geospace environment and has published over 260 papers on these subjects. Dr. Paxton is the PI on seven instruments that have flown in space. Dr. Paxton is an academician member of the International Academy of Astronautics and the past president of the AGU's Space Physics and Aeronomy section. Dr. Paxton's awards include JHU/APL Publication of the Year Awards; JHU/APL Government Purpose Invention of the Year Nominee; and Best Paper – 7th IAA Symposium on Small Satellites for Earth Observation. Other recent relevant experience includes JHU's Global Water Institute and the JHU Earth Environment Sustainability and Health Institute as well as the NASA Heliophysics Roadmap Committee; NSF Aeronomy Review Panel and NSF Aeronomy Committee of Visitors; and chair of IAA Commission 4 and Small Satellite Program Committee. Dr. Paxton earned a Ph.D. in astrophysical, planetary, and atmospheric sciences from the University of Colorado Boulder.

PETE RILEY is vice president, chief financial officer, and senior research scientist at Predictive Science Inc. Dr. Riley is particularly interested in 3D, time-dependent MHD simulations of large-scale heliospheric processes, including solar wind streams and coronal mass ejections. Dr. Riley's expertise lies primarily in developing, testing, and running massively parallel computer codes, which are run on a range of parallel architectures, from small clusters to large supercomputers, such as NSF's Ranger and NASA's Pleiades. Dr. Riley also analyzes a variety of solar and interplanetary data sets and is a member of the STEREO, Ulysses, and ACE plasma instrument teams. Dr. Riley is also a member of the Solar Probe and Solar Orbiter magnetometer instrument teams, was awarded a group achievement award for his contribution to the ACE mission, and co-won the 2006 SAIC Research, Development, Test and Evaluation Group performance award. Dr. Riley has published over 60 papers in the field of space physics, particularly in the area of heliospheric physics. Dr. Riley served as chair for NSF's SHINE (Solar, Heliospheric, and Interplanetary Environment) steering committee and has served on NSF's Space Weather benchmarks steering committee, a follow-on to the NSF Space Weather Benchmarks Phase I study. Additionally, Dr. Riley chaired the 2019 Induced Geo-Electric Fields working group and is also co-lead of the real-time forecasting validation planning group for the Community Coordinated Modeling Center's interplanetary magnetic field Bz, which in turn, is part of the IMF Bz at L1 working team, also co-led by him. Dr. Riley has served as PI for a number of projects supported by NASA, NOAA, NSF, and the Department of Defense. Dr. Riley received a Ph.D. in space physics and astronomy from Rice University.

RONALD E. TURNER is a distinguished analyst with Analytic Services (ANSER) Inc., which in 2004 became the parent institution of the Homeland Security Institute, the only federally funded research and development center dedicated to the Department of Homeland Security. Dr. Turner is an internationally recognized expert in radiation risk management for astronauts, particularly in response to solar storms. For 9 years Dr. Turner was the ANSER point of contact to the NASA Institute for Advanced Concepts, an independent institute charged with creating a vision of future space opportunities to lead NASA into the twenty-first century. Dr. Turner is currently the senior science advisor to the new NASA Innovative Advanced Concepts Program. Dr. Turner was a participating scientist for the Mars Odyssey program. Dr. Turner is on the advisory council to the National Space Biomedical Research Institute Center for Acute Radiation Research. Dr. Turner earned a Ph.D. in physics from The Ohio State University.

NICHOLEEN M. VIALL-KEPKO is a research astrophysicist at NASA's Goddard Space Flight Center and serves as the mission scientist for the Polarimeter to Unify the Corona and Heliosphere mission. Dr. Viall-Kepko studies solar coronal heating, the formation of the solar wind, and the impact of solar wind structures on geospace. Dr. Viall-Kepko has been a highly visible member of NASA's outreach and media team with over 70 live-shot television interviews that included the August 21, 2017, eclipse and the Parker Solar Probe launch and first results. In addition, Dr. Viall-Kepko has given over 100 scientific talks/presentations, 27 international talks, and 39 invited talks. Dr. Viall-Kepko received the 2018 Solar Physics Division of the American Astronomical Society Karen Harvey Prize, awarded for a significant contribution to the study of the Sun early in a person's professional career, as well as NASA's Early Career Achievement Medal in 2018, for "fundamental contributions to understanding coronal heating and the slow solar wind and for valuable service to NASA, the science community and the public." Dr. Viall-Kepko earned a Ph.D. in astronomy from Boston University.

ENDAWOKE YIZENGAW is a senior scientist at the Aerospace Corp. Dr. Yizengaw's research has focused on the complexities of ionospheric electrodynamics, particularly with regard to improving the modeling of the ionosphere as it applies to GPS communications. Previously, Dr. Yizengaw was a senior research scientist at Boston College. Dr. Yizengaw has received the AGU's Joanne Simpson Medal for Mid-Career Scientists. In addition to his scientific contributions, Dr. Yizengaw has played a vital role in the expansion of space science education and research in developing countries, including his native Ethiopia. Dr. Yizengaw holds a Ph.D. in space physics from La Trobe University in Australia.

RAPPORTEUR

ROBERT POOL is a Tallahassee, Florida, based author and editor who specializes in writing about science and technology for a general audience. His most recent book, co-authored with the psychologist Anders Ericsson, is *Peak: Secrets from the New Science of Expertise*, which has now been translated into more than two dozen languages. His previous books include *Eve's Rib: Searching for the Biological Roots of Sex Differences* and *Beyond Engineering: How Society Shapes Technology*, which has been in print for more than 25 years. He has been published in many of the world's leading science magazines, including *Science, Nature, Discover, New Scientist*, and *Technology Review*, and he has worked extensively with the National Academies of Sciences, Engineering, and Medicine.

STAFF

ARTHUR CHARO has been a senior program officer with the Space Studies Board (SSB) since 1995. For most of this time, he has worked with the board's Committee on Earth Science and Applications from Space and the Committee on Solar and Space Physics. He has directed studies resulting in some 38 reports, notably inaugural National Academies' "decadal surveys" in solar and space physics (2002) and Earth science and applications from space (2007). He also served as the study director for the second decadal survey in solar and space physics (2012) and the second Earth science decadal (2018). Dr. Charo received his Ph.D. in experimental atomic and molecular physics in 1981 from Duke University and was a post-doctoral fellow in chemical physics at Harvard University from 1982 to 1985. He then pursued his interests in national security and arms control as a fellow, from 1985 to 1988, at Harvard University's Center for Science and International Affairs. From 1988 to 1995, he worked as a senior analyst and study director in the International Security and Space Program in the Congressional Office of Technology Assessment. In addition to contributing to SSB reports, he is the author of research papers in the field of molecular spectroscopy; reports on arms control and space policy; and the monograph *Continental Air Defense: A Neglected*

Dimension of Strategic Defense (University Press of America, 1990). Dr. Charo is a recipient of a MacArthur Foundation Fellowship in International Security (1985–1987) and a Harvard-Sloan Foundation Fellowship (1987–1988). He was a 1988–1989 American Association for the Advancement of Science Congressional Science Fellow, sponsored by the American Institute of Physics.

ALEXANDER BELLES is a 2022 Christine Mirzayan Science and Technology Policy Graduate Fellow with the SSB. He is a Ph.D. candidate in the Department of Astronomy and Astrophysics at The Pennsylvania State University. His graduate work has focused on panchromatic studies of nearby galaxies and the wavelength-dependent effects of interstellar dust. During his graduate career, Mr. Belles has been a member of the Science Operations Team for the Neil Gehrels Swift Observatory, a NASA space-based observatory with three telescopes used to study gamma-ray bursts. As an undergraduate, he started doing research by studying lithium depletion in open star clusters. Mr. Belles received his B.A. in physics and mathematics from the State University of New York College at Geneseo.

GAYBRIELLE HOLBERT joined the SSB and the Aeronautics and Space Engineering Board as a program assistant in 2019. Prior to joining the National Academies, she was a communication specialist for a non-profit organization that helped inner-city youth by providing after-school programs and resources to engage their needs. Prior to that, she was the social media consultant for the Development Corporation of Columbia Heights and a production assistant for a Startup Multimedia Production Company. She holds a B.A. in mass media communications from the University of the District of Columbia.

COLLEEN N. HARTMAN joined the National Academies in 2018, as director for both the SSB and the Aeronautics and Space Engineering Board. After beginning her government career as a presidential management intern under Ronald Reagan, Dr. Hartman worked on Capitol Hill for House Science and Technology Committee Chairman Don Fuqua, as a senior engineer building spacecraft at NASA Goddard, and as a senior policy analyst at the White House. She has served as planetary division director, deputy associate administrator, and acting associate administrator at NASA's Science Mission Directorate, as deputy assistant administrator at NOAA, and as deputy center director and director of science and exploration at NASA's Goddard Space Flight Center. Dr. Hartman has built and launched scientific balloon payloads, overseen the development of hardware for a variety of Earth-observing spacecraft, and served as NASA program manager for dozens of missions, the most successful of which was the Cosmic Background Explorer (COBE). Data from the COBE spacecraft gained two NASA-sponsored scientists the Nobel Prize in physics in 2006. She also played a pivotal role in developing innovative approaches to powering space probes destined for the solar system's farthest reaches. While at NASA Headquarters, she spearheaded the selection process for the New Horizons probe to Pluto. She helped gain administration and congressional approval for an entirely new class of funded missions that are competitively selected, called New Frontiers, to explore the planets, asteroids, and comets in the solar system. She has several master's degrees and a Ph.D. in physics. Dr. Hartman has received numerous awards, including two prestigious Presidential Rank Awards.